SCALE MODEL ELECTRIC TRAMWAYS

BY THE SAME AUTHOR

British Electric Tramways

A garden scale model tramway scene at Glastonbury illustrating two types of Blackpool tramcars in the foreground with two London *Feltham* type cars behind and a rehabilitated London Class E/3 car in the background, all in ¾ inch to the foot scale

SCALE MODEL ELECTRIC TRAMWAYS -and How to Model Them

E. Jackson-Stevens

With drawings by ERIC THORNTON

DAVID & CHARLES · NEWTON ABBOT

ISBN 0 7153 5480 9

Set in 10/12-point Baskerville
and printed in Great Britain
by W J Holman Limited Dawlish
for David & Charles (Publishers) Limited
South Devon House Newton Abbot Devon

CONTENTS

LIST OF ILLUSTRATIONS

IN TEXT *page*

1 INTRODUCTION

Hobbies in general have never enjoyed so wide a following as they do at the present time and probably the reason lies in increased leisure allied to a degree of affluence which widens the restricted sphere dominant at the beginning of this century. These hobbies are many and varied but rail traction holds a high position, not only in the hearts of its enthusiasts but also in those of the onlookers among the general public which never tires of supporting the increasing number of exhibitions staged throughout the year in cities and towns all over the world.

At one model engineering exhibition in London a national newspaper reporter, alert for a new angle upon which to write, after looking for some time at the wide variety of scale model tramcars gliding over the tracks on the Tramway & Light Railway Society's stand approached me and asked, 'Why model tramcars and not model trains?' 'Simply because on a model tramcar layout there is always movement and action,' I replied. 'In the heyday of electric tramcars it was the slogan of many tramway undertakings that "there's always a tram in sight". While not wishing to decry model railways, when the model train has passed, there should be a scale pause before the next one appears, that is if correct time-keeping realism is to be observed. No such restrictions applied to tramcars, the streets always had a profusion of them.' Duly impressed the reporter left our stand and next day I observed with interest that this theme had been developed to the extent of three-quarters of a column in *The Times* newspaper.

But this is not the only fascination of scale model electric tramcars: we never cease asking ourselves just what are the appeals of this hobby and we find many answers. Nostalgia, perhaps, in seeing miniature replicas of the tramcars which held such appeal in our youth, or possibly the absorption of watching these perfectly scaled little models reproducing on the tracks the motions of their erstwhile prototypes, all under the control of their much-envied operator who is in many cases their owner and creator too. Maybe the appeal of the many continental electric tramways, seen in their efficiency and capacity for moving large crowds, silently and swiftly, has awakened an interest on some holiday, together with the desire to reproduce this form of transport in miniature. Or perhaps the appeal lies in the general unaccountable interest which one never outgrows.

In this hobby, possibly more than in any other, is to be found the opportunity of exercising an exceptionally wide range of talents. Joinery for the baseboards, planning, scheming, drawing, estimating . . . and some pipe-dreaming whilst deciding on the best layout for the tracks. Drilling, pinning, gluing, glazing, and testing while laying the tracks; building, modelling in every conceivable material, overcoming the challenging engineering problems and devising the necessary electrical circuits. Above all, building the tramcars or railcoaches in metal, wood, or any possible new plastic material and glazing them. The scale model tramwayman is architect, builder, engineer, coachpainter, electrician, and many other things besides, each a challenge, each an absorbing new interest, each offering relief within the sphere of the hobby itself when daily routine work becomes tedious.

The choice of scale is naturally determined by the space available but many of us have gardens to maintain and enjoy and there is much to be said for combining two hobbies by building a layout in the wider and more natural setting of the garden. The second consideration is the matter of expense. I know of no similar hobby which provides such an absorbing interest so cheaply. The source of real satisfaction when one is imbued with a desire for models which are different is the building of them one's self, purchasing the more intricate components like motors, controllers, and power units, then machining the trucks and wheels from rough-cast commercial components perhaps with the aid of a friendly lathe-owner. In this pursuit we gain the highest satisfaction of all, that of constructing something of which we may say that it is mainly our own work. In this, we of the scale model tramway fraternity discover the mutual pleasure which arises among people following a common hobby, the mental stimulus of facing and overcoming the problems which arise, and the satisfaction of passing on the lessons learned to the benefit of all.

Now it is sometimes said that we of the scale model tramway fraternity are unhelpful to those on the periphery of the hobby who would like to take a keener interest and perhaps join a society. This is quite untrue and in the many years I have been associated with the hobby, I can affirm without the slightest hesitation that I have never met a more friendly and helpful body of enthusiasts always anxious to share their knowledge and experience, at all times willing to pass on information they have discovered regarding the availability of components and where they can be obtained, and extremely keen to assist a fellow-enthusiast with a problem, whether it entails turning a set of wheels on an available lathe or just painting a crest on a model for those less gifted in this sphere, or any other matter requiring assistance.

One aspect of scale model tramways which often discourages younger prospective modellers at the outset concerns the possession of a lathe. Upon countless occasions at exhibitions I (and other stewards and operators) have been approached by embryonic model tramway engineers clearly apprehensive of the initial cost of a lathe which they think is necessary before they can construct a model tramcar. The whole of this book has been written on the assumption that the tramway modeller for whom it is intended *does not possess a lathe*. It is carefully explained in the following chapters that to purchase a lathe for the sole purpose of making a model tramcar would be an extravagance unwarranted by the extremely occasional items for which it would be needed. I have suggested a number of ways in which the small amount of lathe work can be undertaken and any prospective tramcar modeller with only an elementary amount of enterprise and enthusiasm will not be deterred by not owning a lathe himself. A reliable two-speed electric drill is, of course, essential together with a drill-stand into which it can be inserted for using the grinding-wheel and other attachments.

This, then, may give a brief insight into why we model electric tramcars and tramway systems and derive a considerable amount of enjoyment and pleasure from this absorbing hobby. If some latent interest has been awakened within you, the following pages will provide the benefit of the experience of many tramway modellers.

Here I should like to thank those colleagues in the model tramway field who have so kindly helped with their criticism, advice and the loan of photographs but, above all, my keen appreciation is to Mr Eric Thornton whose renowned scale model tramway drawings have helped innumerable modellers over so many years and who has very kindly undertaken the line drawings for this book. Mr Thornton has also contributed the information in Chapter 15 on smaller scale modelling, in which field he is the widely acknowledged expert.

2 MAIN PRINCIPLES IN THE CONSTRUCTION OF A PROTOTYPE TRAMCAR

Any model engineer, whether experienced or new, devotes keen attention to a detailed study of the real thing before attempting to model it to scale. Not only is research and an appraisal of photographs of the prototype necessary but vastly more important is an appreciation of the detail provided in the scale drawing of the intended model.

Electric tramcars in the British Isles were mainly of two types, single-deck or double-deck cars and there were a number of variants of these two basic classes, four-wheeled cars, eight-wheel or bogie cars, open-topped double-deck ones, or covered-topped examples. There were, however, certain main principles common to the construction of all electric tramcars. The ordinary tramway cars built earlier in this century usually had a wooden body, or superstructure, built on a steel underframe which supported the truck or trucks. In later years steel or aluminium replaced the wooden body. The underframe consisted of longitudinal sole bars formed of rolled-steel angles which supported cross-joists, terminating in channel bars at the ends of the car body. The sides of the cars were completely closed by panelling and glass, the passengers entering and leaving by doors at the ends. The driver's and conductor's platforms were carried on extensions from the sole bars and access to the platforms was gained by steps. The curved dash extended round the front of the car joining the main body on one side only while the collision fenders protruding underneath the dash served to minimise damage to the car in case of accidents.

During and after World War I, it became the practice to glaze the vestibules for protection of the crews and entirely to glaze the upper decks which had previously only had the centre sections enclosed. On the platforms were the brake-handle, a gong-pedal, a pedal for sanding the rails (the sand being stored under seats inside the cars), and the electric controller by which the cars were operated.

The roofs of the cars had to be of sufficient strength to support the trolley-pole which was mounted on a trolley-base, if the current was derived from an overhead trolley-wire. On some systems bow-collectors or pantographs were used for current collection. Tramcar seating on single-deck cars and on the lower decks of double-deck cars was originally longitudinal but in the 1920s was mainly modernised to transverse seating as it always had been on the upper decks, with swing-back seats to enable passengers to face the direction of travel. Sliding doors were provided in the bulkheads of the lower decks and while in some cases access to the central saloon of the upper decks was by the same method, in another variation, a hinged side door on the upper deck gave access to the whole of the upper saloon. Stairways were provided to reach the upper decks at both ends of the car, the earlier type left-handed spirals being changed to right-handed spirals as experience proved the latter type gave the driver an unobstructed left-hand vision and also facilitated loading and unloading.

Interior car floors were protected with wearing strips or slats and the regulations of the Board of Trade (before the advent of the Ministry of Transport) provided that every tramcar should be equipped with an approved pattern life-guard designed to pick up without serious injury any person falling in front of a moving car. The life-guard did not depend on the action of the driver but was entirely automatic. It was known as the 'gate and tray' type and when the 'gate' struck an obstruction on the track, it actuated the 'tray' which dropped on the roadway thus preventing the person knocked down from being injured by the wheels of the car. The tray was re-set by the driver depressing a pedal on the platform.

Tramcars were powered by two electric motors in the case of four-wheeled cars and sometimes two, sometimes four motors for bogie or eight-wheeled cars. The most numerous types of British tramcars were four-wheel cars and the function of the car truck was to form a spring support for the car body and a frame for the driving motors and wheels. The motors were at least partly spring-

Page 17 (above) Actual body underframe of standard type tramcar showing details of construction and method of attaching platform bearers to main body underframe; *(below)* Brill 21E truck constructed by Mr Peter Hammond in $1\frac{1}{2}$in to the ft scale

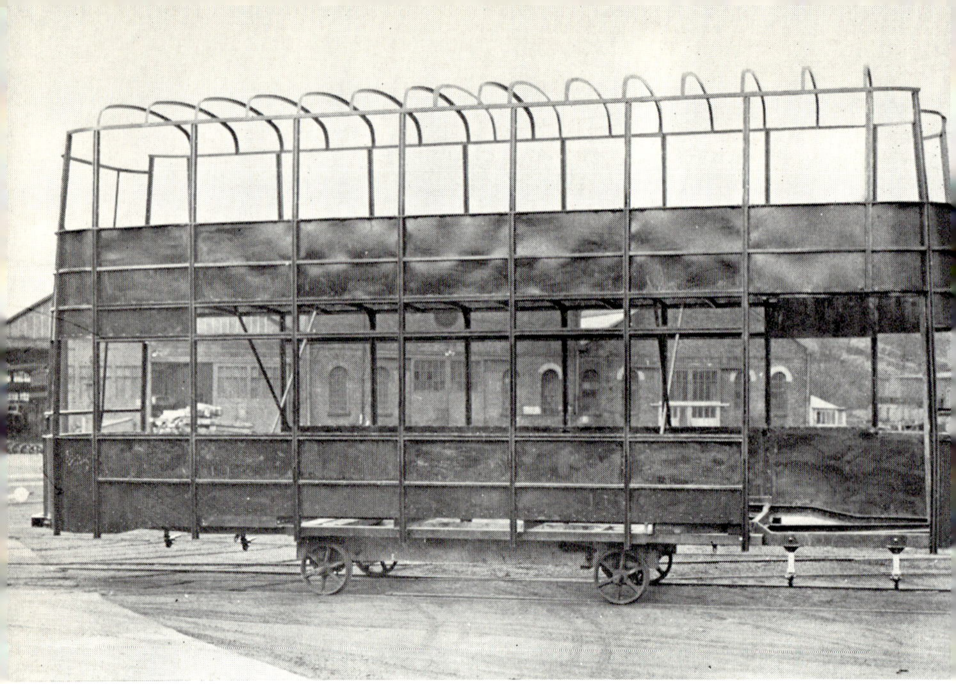

Page 18 (above) Tramcar under construction in makers' yard, on an accommodation truck, showing method of 'framing'; *(below)* a typical prototype reversible-type tramcar seat

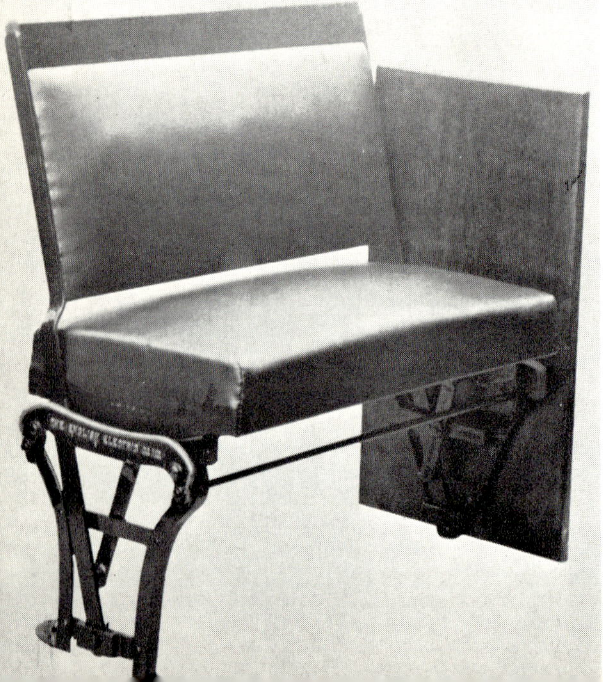

borne and the motor-bearing springs were distinct from the car springs. It was essential that the truck should be easy riding and its construction as uncomplicated as possible in order that the wearing parts could be easily replaced and few in number. Either one or two trucks (if a bogie car) were self-contained, the wheels and axles, motors, driving gear, and brakes all being mounted in one independent framework. The choice of single-truck or bogie car depended on the length of body to be supported. In the case of a single-truck car it was essential for the total wheelbase to be long enough to support the car body without excessive oscillation but the rigid wheelbase must not be too long, otherwise the wheels would bind on curves. Single trucks, with a usual wheelbase of 6ft, were used for car bodies up to about 28ft although, after World War II, some single-truck cars of 32ft were produced, with a correspondingly longer truck wheelbase. A further consideration concerned the gauges, of which there were two main ones in Britain, 3ft 6in and 4ft 8½in. Single-truck cars with a short wheelbase had a greater tendency to oscillation on the 3ft 6in gauges when at speed. Bogie cars were always smoother riding on all gauges, in many cases their wheelbases being 4ft or 4ft 6in each bogie.

The bogie cars in Britain were of two types, equal-wheel bogies and maximum-traction bogies. Equal-wheel bogies were, as the term denotes, equipped with eight wheels of equal size, sometimes with a motor to each axle and, in other cases, with only two motors usually on the outer axles. Maximum-traction bogies had two large driving wheels with two smaller pony wheels. The large driving wheels carried about 70 per cent of the total weight of the car leaving only 30 per cent to be carried by the smaller pony wheels which were not motored. The wheel arrangement of a maximum-traction car was usually Oo—oO although in a few systems the pony wheels were forward thus oO—Oo. On routes with moderate gradients, using one motor per bogie, the maximum-traction truck was better but when heavy gradients were to be climbed the equal-wheel bogie with two motors per truck was the better proposition.

As tramcars weighed anything from 10 to 20 tons, depending on the size of the car, the smooth riding one always expects from a railed vehicle predominated but, having regard to their weight and use amongst other traffic on public highways, exceptionally powerful braking was essential. A detailed study of the different types of tramcar brakes would be too comprehensive for a book of this nature but the main principles can be related. A hand-applied wheel-brake was provided at each end of the car on the platforms, attached to the dash of the car. It consisted of a vertical spindle turned by a brass handle which wound up a chain, under the floor of the car, attached to the brake pull rods. When the handle was turned in a clockwise direction, tightening the chain, the brake rigging was actuated, applying the brake shoes to the wheels. A ratchet was provided by the driver's right foot so that the spindle could be turned with the brake handle always on the most convenient position and where his hand pressure on the handle could most effectively be applied. The brake was kept in the fully applied position, or at any intermediate position (when the car was descending long hills) by means of the pawl which engaged with the ratchet. Brake blocks normally used for braking on the wheels were of chilled cast iron although many different combinations of metals have also been used.

Many tramway systems also used track brakes and, indeed, when the gradients exceeded 1 in 15 over any distance government regulations required these brakes to be fitted. They were known as slipper-brakes and were actuated by a wheel on the platforms of the cars mounted directly underneath the hand-brake lever, the wheel being mounted on a sleeve which passed over the wheel-brake spindle. The wheel was wound on in exactly the same manner as the hand-brake and controlled by a foot-pawl and ratchet likewise but, in this case, the brake blocks mounted over the tracks, between the wheels, were wound down directly on the track rails. The slipper itself was made of oak, beech, or any hard wood.

By far the most effective tramcar braking was that provided by rheostatic braking which acted independently of the power supply to the car, being thus available in the event of any power failure or loss of contact with the power supply, through trolley dewirement or any other cause. This braking was electrically applied through the controller which regulated the supply of current to the motors. Additional contacts changed the connections, allowing the motors to act as generators providing an output which was dissipated wholly or partially through the car starting-resistances thus retarding the car and, if fully applied, stopping it. In addition to this powerful electric braking, on a number of tramway systems particularly in large cities, magnetic track brakes were introduced into the circuit of the ordinary rheostatic brakes. Like the hand-operated slipper brakes previously described, these magnetic track brakes were attached to the car trucks and mounted immediately over the track rails to which they were drawn down with considerable force, when energised upon application by the driver.

Two systems of air brakes were also applied to tramcars of some systems after World War I: the compressed air system and the vacuum system. In the compressed air system the brakes were operated by means of pistons worked by compressed air while in the vacuum system the brakes were applied by withdrawing air from one side of the brake piston and allowing atmospheric pressure to act on the other side. The system of supplying compressed air by motor-driven compressors was the most commonly used. Between the wars a number of the larger tramway systems were equipped with electro-air brakes, a form of track brake in which the slipper or shoe could be operated both mechanically, by means of compressed air, and electrically, as in ordinary magnetic brake systems, the latter being reserved mainly for emergency braking.

Traction motors for electric tramcars had to be substantially constructed because of the heavier duties demanded from them in comparison with stationary motors of the same size. Their design was limited by the fact that the space available for their use was confined to that available beneath the car floor, and between the car wheels. They were designed to be dust-proof and waterproof and so arranged that they could be readily suspended from the truck frames and axles. In the British Isles the size of the motors lay between 25 horsepower and 70 horsepower depending on the local conditions affecting the power necessary to propel the cars and also depending on the type of car. The motors were always geared to the axle. A pinion on the armature shaft engaged with a spur wheel fixed to the axle. The reduction in speed obtained from the motor shaft to the axle depended on the ratio of the numbers of teeth in the spur wheel and pinion. The direct-current motors used for tramways, throughout the world, were adopted because no equally suitable alternating-current motor was available. The maximum voltage supplied to the trolley-wire was fixed by the Board of Trade at 550 volts dc but this voltage was practically standard throughout the world for electric tramways. An interesting story concerning the famous electrical engineer Thomas Edison is that he was once invited to address a gathering of potential electric traction engineers at the Westinghouse works in Pittsburg. Mounting the platform to speak, he pointed to a young man in his audience with the question, 'What is the voltage of an electric tramway trolley-wire?' The young student had to confess he did not know. Neither did any of the students in the room! Edison immediately stumped off the platform with the remark, 'If none of you know the answer to such an elementary question on practically universal standardisation, then I'm wasting my time here!'

The speed of tramway motors had to be varied, and the simplest method of controlling the speed of cars was by means of a resistance used in connection with a switching device known as a controller. Encased in a steel box and operated by a handle on top of the box was a *power-drum* or barrel (so called to distinguish it from the drums used for reversing or braking) provided with copper segments which made electrical connections with fingers when the drum was rotated by means of the handle, thus governing the speed of the car. The power-drum inside the controller had attached to its spindle a star, or index-wheel, which engaged a spring-actuated roller, to compel the power cylinder to take up definite positions corresponding to the various 'steps' on the controller and prevent it remaining in any intermediate position. The colloquial tramway term 'notching-up' describing acceleration arose from the clicking noise made by the star-wheel as the driver increased speed, by moving the controller handle forward, in a clockwise direction from the 'off' position to its full capacity. Likewise, from the 'off' position in an anti-clockwise direction the electric rheostatic and (if also equipped) magnetic track brakes were applied. All controllers were also provided with a detachable key made of brass, about 6 or 8in long, which was inserted near the controller handle and without which the controller could not be operated, or the car started. The key had three positions, 'forward' when it was pointing in that direction, 'reverse' in the opposite direction and 'off' in the intermediate position. The key actuated a drum, much smaller than the power-drum, inside the controller, and its sole function was to immobilise the car and, when the car was needed, free the power-drum to operate the car. All tramway motormen were always under strict instructions never to leave the driving platform without taking the key with them, for obvious reasons.

Board of Trade regulations required that an emergency switch should be provided within convenient reach of the motorman in case of any failure of action of the controller. As tramcars could be driven from either end, these switches were normally provided underneath the canopy over the platform, within his easy reach. The two switches were connected in series in circuit with the main trolley feed into the car. One of the switches had a solenoid and automatic tripping device to open the circuit in the event of excessive current attempting to pass; this is known as the *automatic-circuit-breaker*. The plain switch at the opposite platform is simply termed a canopy-switch. Fuses were also connected in series with the two switches to interrupt the circuit if they failed to do so because, in early days of electric tramcars some cases of failure of circuit-breakers occurred through their not being properly set, with the disastrous result of a controller blowing up. Naturally circuit-breakers were set at a lower interruption capacity than the fuses as it was easier to re-set them than to replace fuses. Lightning arresters were also fitted to all tramcars. Another protective device for open-topped double-deck cars consisted of a trolley-standard-leakage-indicator, in accordance with Board of Trade regulations. This gave warning to the driver or conductor if the trolley-standard became electrically charged because of a fault to earth in the trolley cable. A buzzer sounded and a red light lit up on the platform.

Tramcar lighting circuits included lamps for lighting the interior, conductor's platform, headlight, destination boxes, rear lights, and in the case of covered-top double-deck cars, the upper saloon. As the trolley-wire voltage was 550 volts, the tramcar lights were wired in series in banks of 5 lamps at 110 volts each. The headlight only being required for the direction of travel, and the rear red lights similarly, a two-point switch was provided to reverse this order of lighting at a terminus.

The maintenance and renewal aspect which is sometimes not highly regarded by modellers, provided that all mechanical and electrical equipment was easily accessible. The floors of the lower saloons of tramcars were made to open, granting access to the motors and were known as the 'motor-hatches'. The lifeguards were detachable for easy repair and the brake gear was within easy reach for lubricating and adjusting.

It will have been noticed that phrases like 'normally' and 'generally' have been used. This is simply because it is impossible to be dogmatic where electric tramcars and their equipment is concerned as variants occurred, particularly as modifications and improvements took place in the light of experience throughout the years. The principles exemplified were, in the main, adhered to and should form a reliable basis for tramcar modelling.

3 DECIDING ON THE SCALE IN RELATION TO THE PROTOTYPE

Upon this question depends the success of the final model and the layout on which it is intended to run it. There are many individual variations and a wide choice is available. The ingenuity of many modellers in the smaller gauges is admirable and the minute detail attained often commendable but the fact does remain that, in some cases, when only the tramcar body has been constructed and fitted to a motored truck purchased from a manufacturer, this does not constitute model engineering in the strictest sense of the term. At the same time, a number of tramway modellers have successfully constructed their own trucks and fitted minute gearing and motors to them in the diminutive 4mm and 7mm scales.

The scale to be decided upon sometimes depends on the room available for the eventual layout and also upon the workshop facilities at the disposal of the modeller. Here again, the modeller will have to decide whether he intends to build a substantial model in which the proportions will allow the considerable detail of the prototype to be copied in impressive thoroughness or whether he is more concerned with building an eventual fleet of smaller tramcars to be operated on a restricted indoor system. If he decides on a substantial model, the question will arise regarding where it is to be run. In the larger scales, the proportions of the model do not permit indoor running and a garden layout is the only possibility unless, of course, the modeller lives near an established model tramway club which has a layout for the use of its members or alternatively, if the modeller builds his model with the intention of only running it at exhibitions where space is available.

In the smaller scales it is possible to equip a room or loft in a house with the necessary track and overhead system but a problem which should not be overlooked arises when the necessary hinged 'swingbridge' to allow for the door giving access to the room has to be considered. This is not insurmountable as there are a number of examples of bascule bridges in Holland where the problem of successfully raising the track and the overhead trolley-wires in one operation has been overcome. But the decision on the scale to be adopted merits most careful thought beforehand and the following table of standard scales and gauges will help in deciding:

Scale to the ft	Rail Gauge	Radius of Curve				Average Maximum Speed
		Minimum		Average		
		ft	in	ft	in	mph
4mm	No 00 16·5mm	1	9	2	0	$\frac{1}{2}$
7mm	No 0 $1\frac{1}{4}$in	2	0	3	6	$1\frac{1}{4}$
10mm	No 1 $1\frac{3}{4}$in	3	6	4	6	3
13·5mm (17/32in) 'Half inch'	$2\frac{1}{2}$in	5	0	8	6	5
$\frac{3}{4}$in	$3\frac{1}{2}$in	10	0	16	0	7
1in	5in	15	0	25	0	9
$1\frac{1}{2}$in	$7\frac{1}{4}$in	35	0	60	0	12
3in	15in	100	0	150	0	30

There are variations between the above standard scales but only the most commonly used have been given as any departure from the standards laid down produces difficulties in finding suitable motors and gearing. The first essential is a drawing of the car which it is intended to model, to the scale chosen. Drawings can be obtained but it may be necessary to scale them up or down to the scale required; another essential requisite is a good clear photograph, or photographs from several angles, of the prototype of the projected model. Some modellers build the bodywork first but the majority construct the truck or trucks initially and the body of the car when these are completed.

The smallest 'TT' gauge is only suitable for a table-top and, although some modellers do adopt this tiny scale and also OO gauge, neither could be regarded as actual scale modelling in its truest sense as both scales rely upon the purchase of ready-made kits of parts, anathema to the purist scale modeller. Quite true, pleasure of operation can be obtained from these diminutive cast-metal kits but they are always troublesome through excessive derailments and 'sticking' or failing to start when power is applied. Nothing looks worse, at exhibitions, than to see operators having to give models a push to make them start and the general public attending model exhibitions are not impressed by the erratic control of these undersized models. Not only do they spring into action with the alacrity of a grasshopper but they stop with equal abruptness giving one the impression that, if this is how prototype electric tramcars behaved, then the passengers must all have been thrown in a heap against the bulkhead when the cars braked. Actual scale modelling takes into account every aspect of the prototype which can be reproduced accurately and this includes acceleration, braking, and running speed. The distinction between a child's toy 'model' running round a track at a speed which would represent 200mph and a well-built, properly controlled scale model is too obvious to merit any comment. Even so, 4mm or OO gauge does not produce any satisfaction for the dedicated scale-modeller.

In 7mm (O gauge) and 10mm (1 gauge), however, some very realistic models have been made especially when space limitations are a consideration, for many enthusiasts do model in these scales and they have quite a following. The previously described erratic starting and stopping has been completely overcome by equipping the motor with a flywheel thus enabling the model gently to coast to a stop exactly as in prototype practice. Chapter 15 is devoted to every aspect of modelling in these smaller scales. A number of models in 13·5mm scale have been constructed in some areas and one or two isolated instances have occurred of tramcars being made in 1in and 1½in scales. The 3in scale is designed for passenger-carrying models and an extensive garden or orchard would be required for this.

By far the most popular scale is the 3/4in to the foot adopted by the Tramway & Light Railway Society as standard, for modellers who are concerned with detail to an impressive degree. This scale was decided upon by the founders of the Society in 1938 as being the smallest practicable scale in which the prototype tramcars could be reproduced with a lifelike exactitude. An overriding consideration in the choice of this scale was the fact that this is the smallest scale in which it is possible to mount the driving-motors *underneath the floorboards* of the tramcars at the same time retaining correct scale motors. This principle applies when the prototype is to either of the two main gauges used by tramways: 4ft 8½in (which scales down to 3½in model gauge) or 3ft 6in (which scales down to 2⅝in model gauge). The continental metre-gauge of 3ft 3in will also scale down accurately in 3/4in to the foot scale. In all other scales from the smallest up to this scale of 3/4in, it is necessary for the motors to protrude into the lower saloon of the tramcar thus preventing accurate modelling of the floors, which is not realistic or true to type.

The standard gauges of 4ft 8½in used mainly in England, America, and Europe provide many worthwhile prototype tramcars of appeal to modellers while the 3ft 6in gauge used in a number of English cities can also be modelled in this 3/4in scale as 3/4in does not affect the scale, only the gauge must be 2⅝in. Indeed, many fine Birmingham, Llandudno, Wolverhampton, and other similar gauged 3ft 6in tramcars have been constructed in this scale, as have a number of continental metre-gauge models.

Probably an overriding consideration in the choice of scale is whether the prospective modeller intends to achieve the acme of success by ultimately running his model at one of the many model exhibitions, with other modellers who exhibit their models annually. If this is his goal, then 3/4in scale is the main choice as practically all exhibition portable track is to this scale, in both gauges, and exhibition portable track is made solely for this purpose and owned by a Tramway Society. There is, of course, nothing to prevent a modeller constructing his own portable track, in addition to his fixed track at home, and taking it to an exhibition, but this is an expensive alternative. Such an expense is hardly justified when a Tramway Society portable track and all the relative overhead equipment is supplied for members. It is also possible for a modeller to exhibit his completed model as a static model on a small length of track just sufficient to support it but this hardly has the same appeal as a working model.

If the prospective modeller intends to construct an outdoor scale model tramway then anything less than 3/4in to the foot scale will present him with many problems discussed in Chapter 4.

4 THE TRUCKS

If there is one aspect of tramcar modelling which has done more than any other to deter the average amateur engineer from making a start in constructing a model tramcar it is his doubts regarding his capacity to construct a truck. Emanating from America, the word 'truck' is used to describe the self-contained assembly upon which the wheels, axles, and motors are mounted. In early days, on some systems, an eight-wheeled or 'bogie' tramcar was referred to as a 'double-truck' car to distinguish it from a 'single-truck' or four-wheeled car.

The first consideration which will cause the prospective tramcar modeller some heart-searching is whether he will have to purchase a lathe, an expensive investment, before he commences his project of making a scale model tramcar. Twenty years ago the possession of one's own lathe would have been an essential first step but, with the growth of metalwork classes for apprentices and others and the help of co-operative friends in engineering works for the small amount of lathe-work required, in comparison to the whole project, the purchase of a lathe especially to make one, or several, model tramcars would be an excessive luxury. Moreover, the availability of castings and material now obtainable simplifies the construction of tramcar trucks.

The most commonly used tramcar trucks were those exemplified in the castings drawings opposite (Fig 1). They are all obtainable from the Tramway & Light Railway Society and have been a boon to members of that Society in the many models they have constructed. Sturdily cast in bronze they require very little smoothing and finishing and they are to exact $\frac{3}{4}$in to the ft scale measurements. If non-standard types are required, the modeller will have to cut out and machine these himself. However, it is well to bear

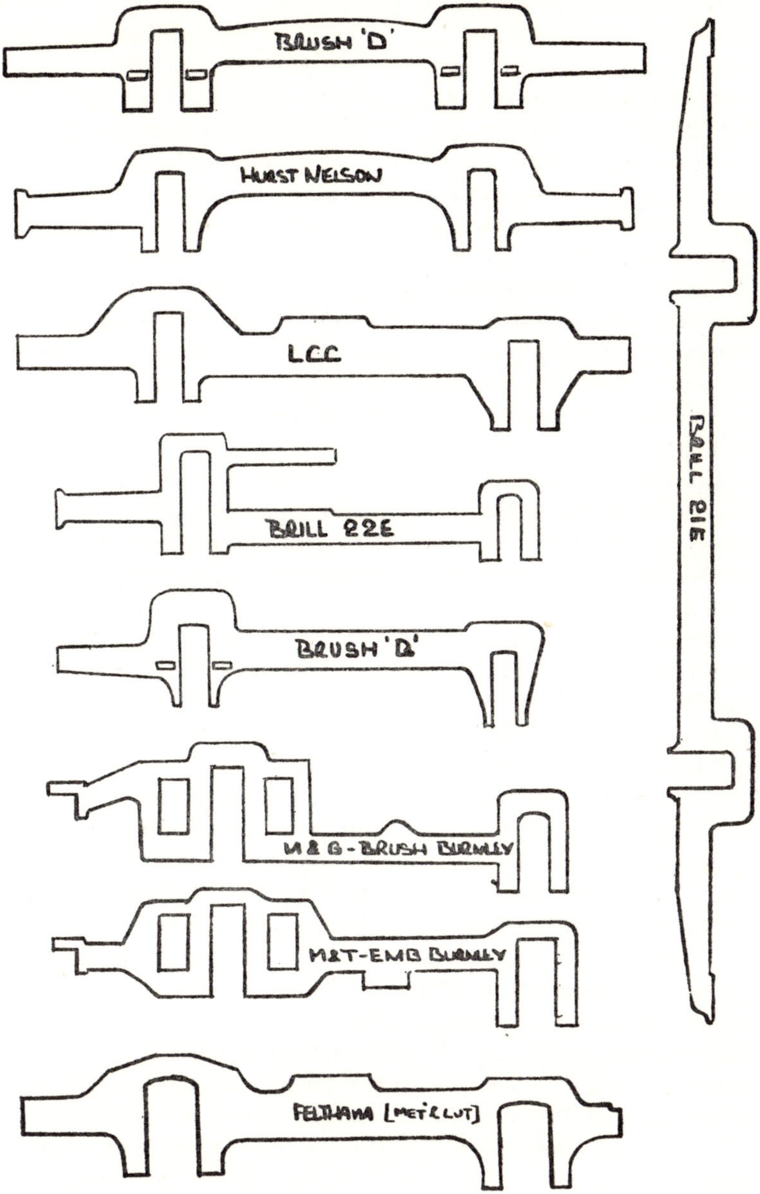

FIG 1 Bronze truck and bogie side frame castings (courtesy of the Tramway & Light Railway Society)

in mind that the accurate castings shown also include the detailed moulding of the original prototypes. But the purchase of the necessary truck side-frames is only the first step and does not, by itself, ensure that the completed model is well on the way. Indeed, the late Mr Frank J. Roche, the former chief engineer of the renowned modelmakers, Bassett-Lowke Ltd, once told me that there were more side-frame castings lying around in chests of drawers up and down the country than any other components! Impressed by commendable models which are seen at exhibitions, on the spur of the moment a prospective modeller determines to 'have a go' and purchases the side-frames as an earnest of his intention. Unfortunately when he gets home, away from the sight of the glistening rolling-stock of the exhibition, his inspiration wanes and his determination fades. This is a pity because making a model is a challenge and a worthwhile project.

Although for the tramcar modeller a lathe is not vitally essential a good electric drill certainly is and, together with a horizontal drill-stand and the relative ancillary drill equipment allied to a reliable electric soldering-iron, considerable accurate progress is possible.

One of the most popular four-wheeled trucks in use for electric tramcars was the Brill 21E illustrated in Fig 2. The name Brill was that of the patentees, J. G. Brill & Company of Philadelphia, USA, but they were also made under licence in England. This is by far the easiest type of truck for any tramcar modeller to attempt in his first model; its construction is quite straightforward. A scale drawing, obtainable from the Tramway & Light Railway Society is, of course, essential. The underframe of the car body will be bolted to the top plates *c* of the truck so that the weight is supported and transmitted to the side-frames *a* by four helical springs *d* and two elliptic springs *e* on each side of the truck. The helical springs *d* maintain the principal dead weight of the car while the elliptic springs *e* prevent the oscillation of the car. The main side-frames *a* of the truck, carrying the whole weight of the body and truck, rest on the helical axlebox springs *f* of which there are four on each side of the truck. These springs *f* rest on wings, or brackets carried out from the sides of the axleboxes *g* which are themselves hung directly on the car axles and are free to slide vertically in guides formed by the vertical limbs of the side-frames at these points. The whole side-frame of the truck is maintained rigid by

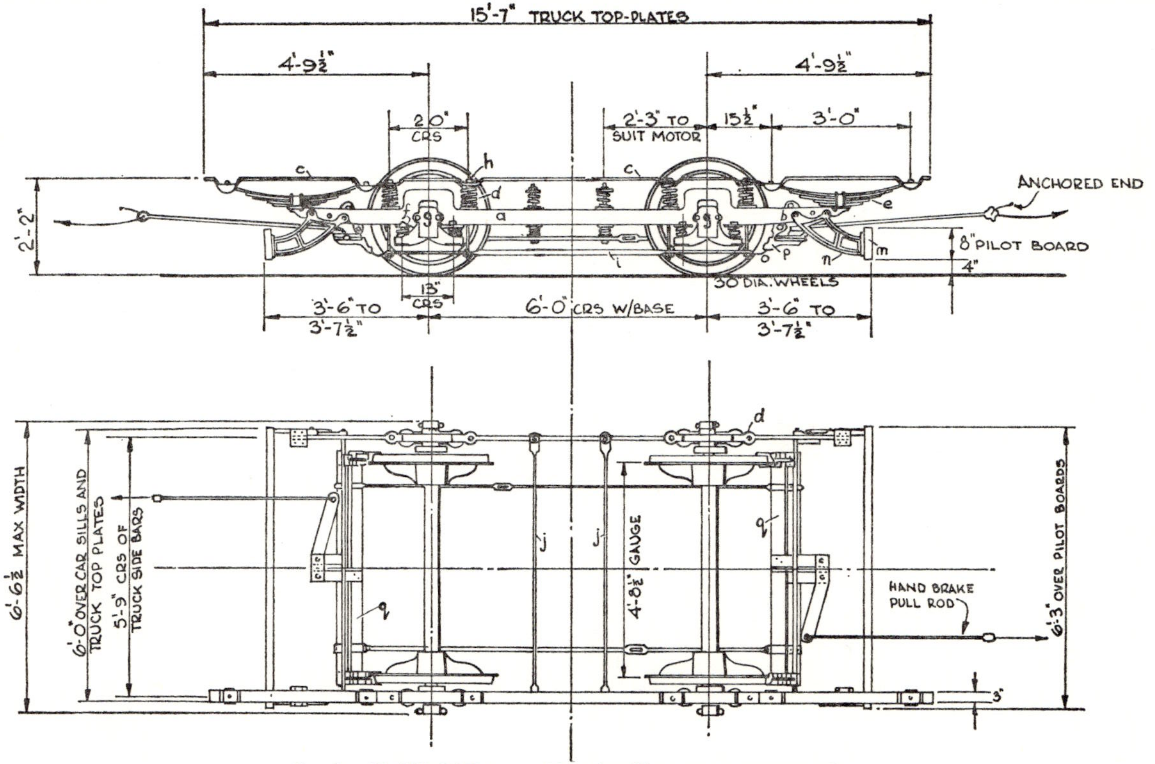

FIG 2 'Brill' 21E type 6ft wheelbase tramcar truck

taking the spring posts *h* of the body springs *d* through suitable holes in the side-frames and by staying the spring posts together by means of the stay *i*.

The side-frames of the truck receive lateral support by the diagonal brace bars *j* from which the motor supports are spring-supported. Two motors are used, one driving each axle. The weight of each motor is divided between the axle and motor-beam from which it hangs. The pilot-boards *m* are attached to side bars at each end of the truck by pilot-board-brackets *n*. The pilot-boards are used for the attachment of the life-guards. The brake shoes *o* are held in holders *p* fixed to the ends of the transverse bars *q* or brake beams; these are hung from the truck end bars *b* by the brake-block hangers. A rectangular frame rests on the brake beams and slides on them when the brakes are applied by winding the brake-lever. The brake blocks are kept clear of the wheels by means of the brake-release springs and are, therefore, only in contact with the wheels when the brakes are applied.

The modeller will need to experiment with the small helical springs needed for correct tension, having regard to the weight of the finished model. Springs too hard cause rough riding and derailments while weaker springs will give excessive oscillation. The same applies to the four elliptic springs which are easily made from phosphor-bronze strip. Before attempting to construct a truck, it is essential for the modeller to lay in a selection of all ranges of 6BA to 12BA steel and brass nuts and bolts, hexagon head and cheese head, plain washers and, of course, the relative drills. Some of these latter can be obtained from the local ironmonger but the nuts and bolts will have to be bought from a model engineering supplier, as will the necessary lengths of brass bar, strip, and channel. It will also be necessary to obtain a short length of brass bar for the axleboxes as in 3/4in scale these are 3/4in high by 1/4in deep while the wings carried out from the sides of the axleboxes extend to 1½in and it is advisable to cut the four axleboxes from solid brass bar. It is better to overstock with model engineering nuts and bolts than otherwise as nothing is more frustrating than running out of a vital correct-sized nut or bolt, or both, when working on a model and the supplier may be a long distance away.

The construction of scale model tramcar bogie trucks differs in certain respects from that of four-wheel 'single-trucks' as bogie

Page 35 (above) Blackpool double-deck bogie car under construction in 3/4in to the ft scale, showing the lower saloon at the top, and below, the brass 'shell' of the same model with the upper saloon added. The lower saloon has already been glazed, as can be seen by the whitewash on the windows; *(right)* London County Council 'M' class car no 1430, constructed by Mr E. R. Oakley in 3/4in to the ft scale, shown running at a London exhibition. These four-wheel cars worked on routes where traffic was light and were distinguished by being equipped with a special truck designed by Hurst Nelson & Co in conjunction with the LCC—this can be clearly seen on the model

Page 36 (above) Mr Robert Whetstone's Birmingham Corporation open-top bogie tramcar No 2 in $1\frac{1}{2}$in to the ft scale, shown prior to assembly; *(below)* 3/4in to the ft scale model of Sunderland Corporation single-deck bogie tramcar No 85, constructed by Mr Gordon Douglas

trucks are made to swivel and, therefore, some of the principles of springing, previously described, do not apply and another form of springing is used. Bogie trucks are mainly 'swing-bolster trucks' although some were constructed without swing bolsters. As the swing bolster principle was widely used, the method of construction is given. Two types of bogie trucks were widely used by electric tramways, the 'equal-wheel' trucks in which, as the name implies, all wheels were of equal size and the 'maximum-traction' type trucks where the driving wheels were of standard size while the pony wheels were smaller. The principles involved are explained in Chapter 1 dealing with the prototypes.

A typical equal-wheel centre-bearing swing-bolster truck is shown in the illustration, Fig 3. It will be noticed that the side-

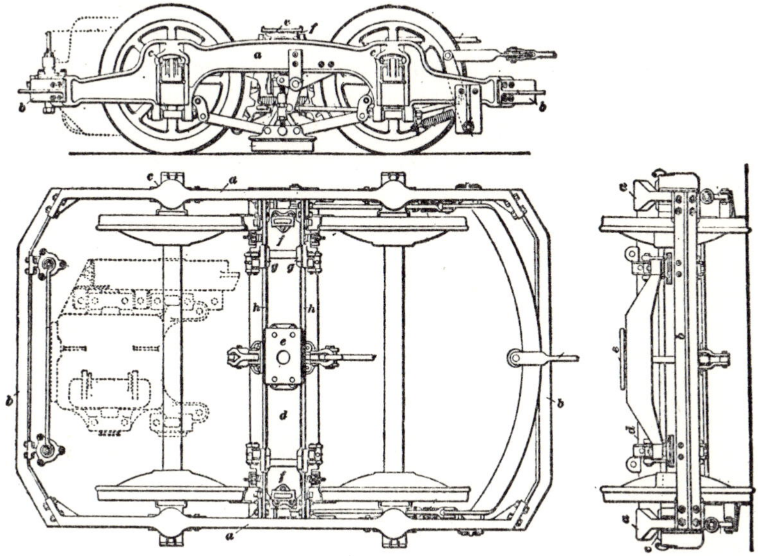

FIG 3 Equal-wheel bogie

frames of a bogie-truck are smaller in length than a single-truck for the reason that the car weight is distributed over eight wheels with a correspondingly longer body in bogie-cars. The side-frames *a* are joined by the end cross-bars *b*. Helical axlebox springs *c* on the top of the axleboxes fit into cups machined out of the underside of the side-frames. The bolster *d* supports the car body on the swivel plate *e* and rubbing plates *f*. The centre pin fits into the centre hole at *e*. The ends of the bolster are spring-supported on helical springs bearing on a spring plank which is suspended by links *g* from transoms *h* which cross the side-frames of the truck. The track-brake shoes are applied to the rails either magnetically or through the hand-brake linkage previously described.

In the case of a maximum-traction truck the car body rests on the top of a swivel plate which is pivoted to the lower swivel plate hinged to the bolster in order that the swivel plates may rock slightly thus equalizing the load distribution on them. Additional rubbing plates are bolted to each end of the bolster to balance the car body. (Fig 4.)

The axleboxes in model tramcar construction can be machined to pattern out of brass bar. Normally the oiler blocks, felt pads conveying oil to the journals, oil reservoirs, etc, a feature of the prototype tramcars, are not reproduced in a model of 3/4in to the foot scale. The axlebox is drilled to take the axle-journal and the top of the axlebox should have a slightly recessed seat for the axle-box truck springs to rest in as these are helical. For the purposes of modelling, to save machining out the prototype axlebox cover and fitting a cover, all that is necessary is to drill a small hole through the centre of the spring-seating on the top of the axle-box as far down as the journal, for the occasional insertion of penetrating oil.

The types of trucks described by no means exhaust the possibilities although they exemplify the main principles. A variety of trucks were employed in England, America, and throughout the world. Manchester, for instance, had a preference for Brill 22E bogie maximum-traction trucks up to about 1921, but after this they designed a maximum-traction truck themselves, based on the Brill 22E design but having an extra top bar for additional strength. Thirty-eight tramcars of an entirely new design were also produced by Manchester incorporating some remarkable features in four-wheelers. Known as the 'Pilchers' (named after the general manager, Mr R. Stuart Pilcher, who introduced the design) their trucks were a modification of the Peckham Pendulum P35 truck, made by the Brush Company at Loughborough under patents acquired from the Peckham Car Wheel Company of America. This pendulum gear was a device to allow the axle some slight freedom of lateral movement in relation to the truck frame to

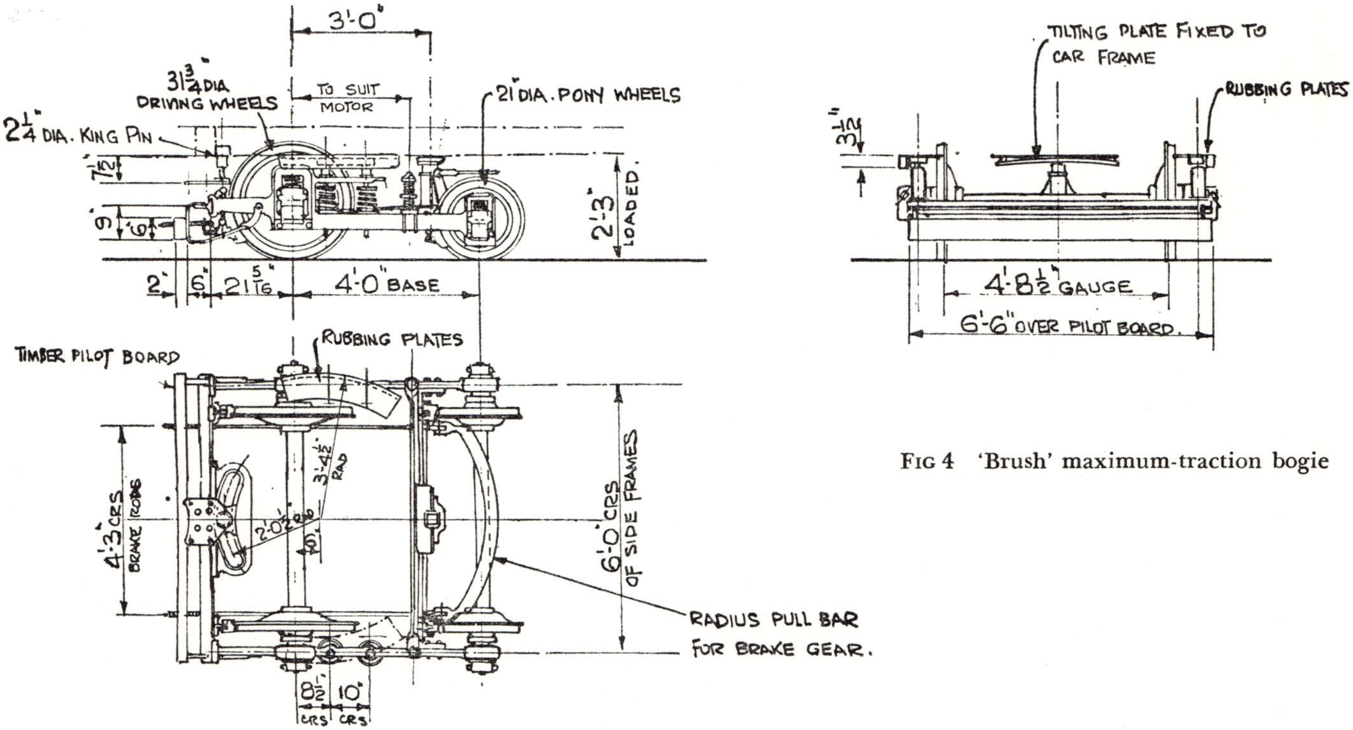

FIG 4 'Brush' maximum-traction bogie

minimise the effects of track irregularities, and the distance between the axles was 8ft 6in. Birmingham had a number of Brush Burnley bogies and Maley & Taunton Burnley bogies while London County Council used Mountain & Gibson, McGuire, Hurst Nelson, EMB, and some Brill 22E bogie-trucks. On their 'M' class four-wheelers, the LCC used a specially designed Hurst Nelson type of truck. A variety of trucks were used by Glasgow while in later days a number of inside-bearing Maley & Taunton ones were used. Their 'Green Goddess' cars purchased from Liverpool Corporation Transport (who built them in 1936-7) were mounted on Maley & Taunton swing-link bogies. From the examples given it will be seen that in the research undertaken before the modelling of a tramcar is put in hand, it is essential to ascertain the correct type of truck(s) used.

If the modeller has in mind a more ambitious model, on the lines of the streamlined railcoaches now used in America and extensively on the Continent of Europe, a resumé of their history to the present day may be helpful. Due to the necessity to modernise the American tramways the chiefs of the main tramway undertakings and tramcar builders formed the Electric Railway Presidents' Conference Committee in 1930. The object of the Committee, consisting of the chiefs of twenty-five undertakings and builders of tramcars, was to carry out research into designing an entirely new and standardised tramcar. Eventually a revolutionary prototype, known as the PCC (Presidents Conference Committee) incorporating many startling new features, was produced in 1934. The trucks of this car were equipped with lightweight motors which drove through a cardan shaft and worm gearing. Rubber-cushioned wheels and rubber-assisted suspension added to the comfort of the riding of the car which, with the new welded track, provided a noiseless vehicle.

The term PCC tramcar became the accepted term for describing cars of other countries copying this innovation. PCC cars are operating in many cities of the Continent and in the Communist countries, where their adoption has been widespread. The only PCC type tramcars in Great Britain are those built for Blackpool Corporation on Maley-Taunton hornless, equal-wheel, cardan-shaft-drive bogies specially designed for high speed running and fitted with resilient wheels. These wheels contain rubber pads which prevent the transmission of high frequency vibrations from the wheel rims to the trucks and thereby increase the life of the vehicle as well as improving passenger comfort. Truck wear is reduced and a notable decrease in noise is effected. The trucks are of the side bearing type and smooth riding is given by a combination of steel semi-elliptic springs and rubber springs controlled by dampers. There are two motors on each bogie, driving through propellor shafts and spiral bevel gears. In addition to track-brakes, wheel-braking is on brake drums extended outside the running wheels, as it is not possible to brake on resilient wheels. On some similar trucks built in Europe, braking is effected on the axles in some cases. An amateur modeller attempting to model trucks of this advanced design would certainly need access to a lathe and considerable engineering knowledge.

5 TRAMCAR BRAKES AND THE MODELLER

As all scale model tramcars are remote controlled the consideration of equipping the models with operative brakes is a matter for the individual modeller to decide for himself. Many tramcar modellers are content with fitting non-operative brakes for appearance sake but, for the perfectionist, a description of prototype tramcar braking is given in so far as it would be practicable to model it.

When any tramcar, or railed vehicle, is running on the level and the power is shut off, it will continue to run for some time because of the kinetic energy stored in it but, due to the friction of the wheels and axles and air resistance, it will eventually stop. This principle is relied on in operating model tramcars and, with care in controller operation, it has been found effective on extensive layouts and also in the confined space layouts at exhibitions.

But in actual prototype tramcar practice it is necessary to stop the cars in shorter distances and less time by means of brakes. Unless the modeller operating the model has an assistant running alongside the model tramcar it is difficult to see how this principle could be applied in model practice! However, the types of brakes used on tramcars are mechanical, electrical, and compressed air, or combinations of these. A simple form of common single-truck brake rigging is shown in Fig 5. The brake beams are supported at the ends where they slide in guides bolted to the side-frames of the truck. The equaliser rods are rigidly connected to the equaliser bars in which the brake beams can slide freely. The levers underneath the equalising bar transmit pull on the pull-rods or brake chains to draw the brake beams towards each other and apply the brake shoes to the wheels. In the illustration the brakes are shown

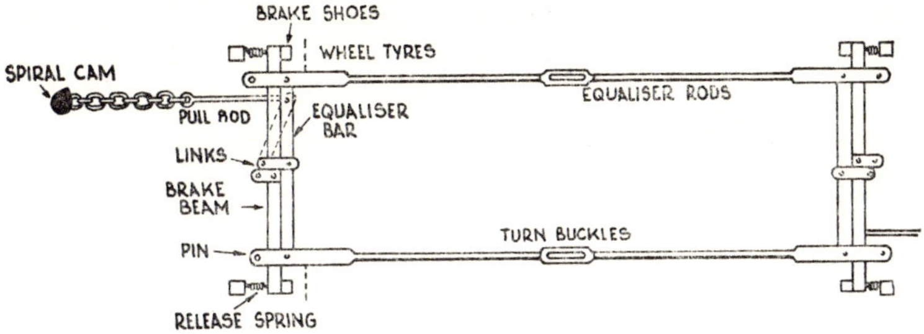

FIG 5 General principles of single-truck handbrake rigging

in the off position and the shoes have been pulled away from the wheels by the release springs. One end of each release spring is attached to the car truck and the other to the brake beam. The turnbuckles on the equaliser rods are adjustable in order to take up any slack and adjust the brake blocks.

The materials required for the construction of scale model tramcar brakes and brake rigging in 3/4in to the foot scale are brass strip 1/8in by 1/16in, brass rodding 3/32 and also 1/8in in diameter, the small release springs, a quantity of 8 BA brass nuts and bolts, a small section of brake chain and a section of 3/8 by 1/4in brass bar, this latter from which to machine the brake blocks.

The illustration (Fig 6) shows a type of double or bogie truck brake rigging in common practice. In order that the rear brakes are not applied harder than the forward ones an extra equalising lever that cannot easily be used in single truck riggings is provided. The pull on either of the pull-rods is transmitted through the brake levers and tension rods to the arch bars on each truck, enabling the rollers on the tension rods running on the arch bars to apply the brakes whatever the positions of the bogies. The illustration shows the position of the best leverage, the brake blocks being on the wheels.

The truck rigging on one side of the truck is shown and it will be seen that a pull to the right on the pull rod forces the brake shoes against their respective wheels as the fixed point to the car truck is held in place by a bolt and nut.

Although the brake rigging for maximum-traction bogies is the same as for equal-traction bogies the truck rigging is modified so

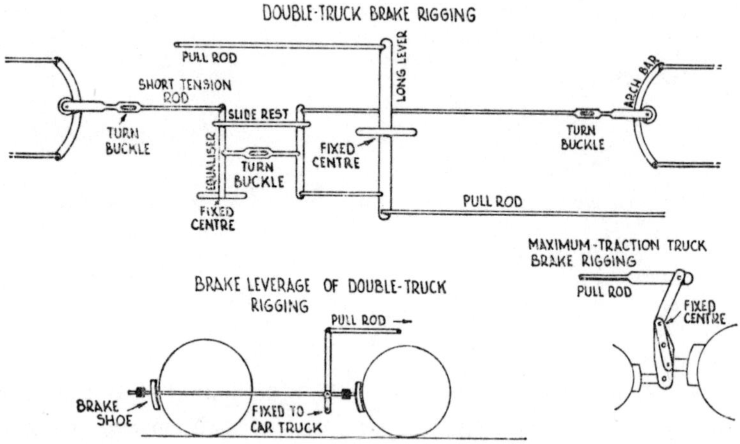

FIG 6 Bogie brake rigging

that the greater braking pressure is applied to the larger wheels, as two-thirds of the weight of the car is on these wheels. The truck rigging shown in Fig 6 shows the truck pull-rod connected to the top of a bent lever. The lower end of this lever is attached to a push rod which operates the brake shoe of the large wheel while the extreme base of the lever operates a push rod applying the

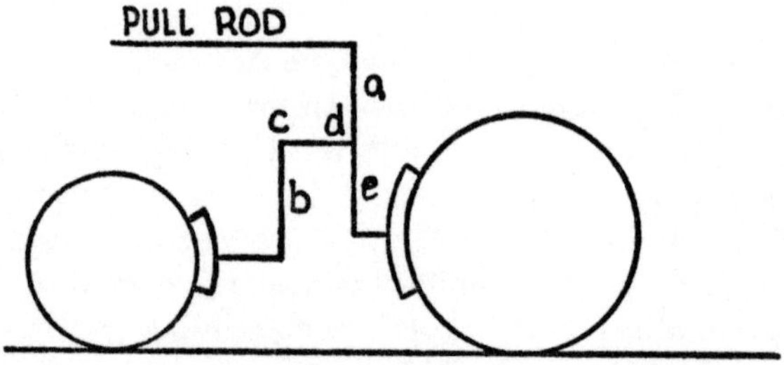

FIG 7 Diagrammatic principle of operation of maximum-traction brakes

smaller brake shoe to the pony, or small wheel. The movements of the large and small shoes take place simultaneously. The principle of maximum-traction truck braking is really quite simple and is shown diagrammatically in the illustration. When the rod *d* becomes short the connection between the two levers becomes coincident at the actual pivot centre of *d*. Lever *b* then becomes an equalising lever for proportioning the brake-block pressures. The truck pull rod is connected to the upper end of lever *a* and *e* which is connected to the pivoted *d* which has a stationary fulcrum at *c* thus applying greater pressure to the large wheels and correspondingly smaller pressure to the smaller wheels.

Tramcar handbrakes are normally constructed to operate by winding the handbrake handle in a clockwise direction which in turn rotates the brake stock. This stock is attached to the dash of the car by brackets and when the handle is turned it winds up the chain underneath the platform, pulling on the brake rods. The Peacock-type brake handle shown in Fig 8 causes the stock *b* to

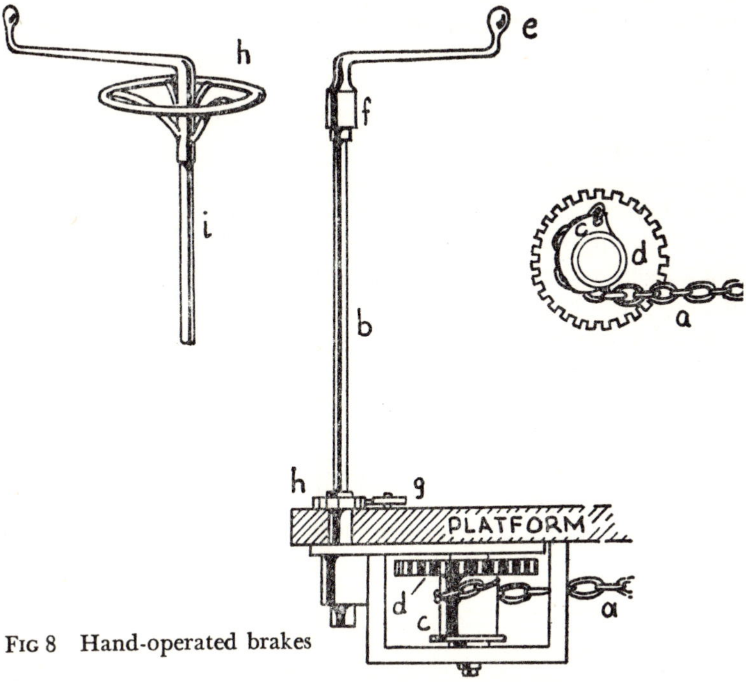

FIG 8 Hand-operated brakes

wind the chain on a spiral cam *c* grooved to receive it, and fixed to a larger toothed wheel *d*. This cam is so shaped that when the brake handle *e* is turned the slack of the chain is taken up. A ratchet is provided in the handle *f* to turn the brake handle in the most convenient position. The brake can be held on in the full position or at any intermediate position by means of the pawl *g* which operated by the driver's foot engages a ratchet wheel *h*.

In addition to the wheel brakes previously described, some tramway systems were required by law to equip their cars with track brakes in hilly districts. These brakes were also known as slipper brakes and were applied directly to the tracks, by a handwheel mounted on a sleeve which being outside the normal handbrake stock revolved independently of the handbrake (*h* and *i*). The sleeve actuated a pinion and a pair of bevel gears underneath the platform to which the pull rod of the track brake was attached thus applying the track brake to the rail. The track brake handwheel spindle is controlled by a pawl and ratchet, slightly to the right of the handbrake one, in exactly the same way as the handbrake itself. For modelling this handwheel the brass rod should be carefully shaped and soldered on to the sleeve which is made from brass tube. This latter should not fit too tightly on to the internal handbrake stock as it has to revolve independently.

Rheostatic braking was used on many of the larger systems of England and is widely used on the Continent and in America. Some isolated modellers have, with considerable patience and the ingenious application of electrical knowledge, succeeded in making a modification of this system work with their remote controlled systems. The principle governing rheostatic braking is that the controller regulating the supply of current to the motors is provided with additional notches which change the connections to the motors in such a manner that they act as generators back to the resistances. The connections for rheostatic braking are illustrated

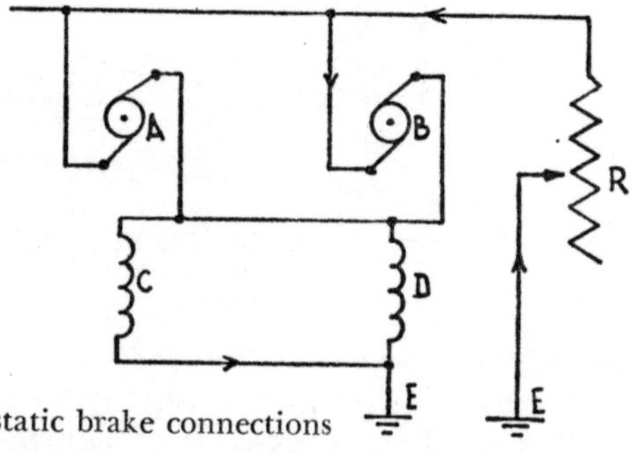

FIG 9 Rheostatic brake connections

in Fig 9. The back electromotive force of a motor opposes the flow of current through the armature from the line. If the motor is disconnected from the line and it continues revolving the back emf will set up a current in the opposite direction to the original motor current. The resulting magnetism will tend to send the current in the opposite direction to the original exciting current. By closing the field circuit through the motor and a resistance, the motor field will not build up as a generator, becoming demagnetized altogether. To make a series motor act as a generator either the field or armature connections require reversing. It is usual to reverse the armature connections. The two motors are connected in parallel to each other with the connections at the armatures *A* and *B* reversed and in series with the car starting resistance *R* through the earth wire *E* of the car. An equalizing connection across the terminals of the field windings *C* and *D* next to the armatures ensures that the field strength and currents of the two motors are equal. The foregoing does not, of course, apply to permanent-magnet motors, which have no field windings.

But whether the modeller attempts to try a rheostatic braking circuit on his layout or not, he will have to give some attention to an electromagnetic track brake as these brakes were used as a service brake on the larger systems. One of the most common types of magnetic track brake was the Westinghouse one illustrated in Fig 10. A magnetic track brake must not be confused with the

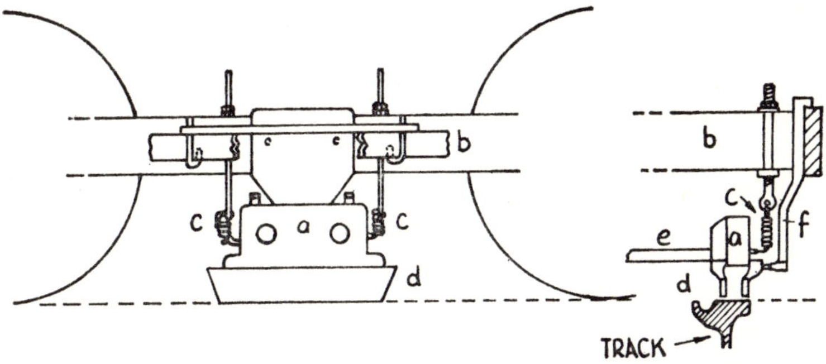

FIG 10 Westinghouse electromagnetic track brakes

purely mechanical track brake previously described. The Westinghouse magnetic track brake relied for its operation upon its introduction into the circuit of the ordinary rheostatic brake. Designed for attaching to the car trucks and mounted directly over the track rails, they are drawn down magnetically with a force controlled by the graduated brake application on the controller. Two brake magnets are required for a single-truck car and four for a bogie car. The current generated for rheostatic braking is passed through the exciting coils wound round the cores of the brake magnets. The illustration shows a side view and a transverse view of the braking arrangement. Each brake magnet *a* is suspended from the truck frame *b* by flexible springs *c* in such a manner that the magnet is normally suspended clear of the rail *d* when not in operation. Opposite magnets are connected by a crossbar *e*. The retardation of the brake magnet on the rail is transmitted to the truck frame through the tongue-shaped bracket *f* bolted to the truck.

As modification in modelling is quite usual, the normal practice is to construct the magnetic track brakes and compromise by leaving them non-operative. The tongue-shaped bracket *f* is made from 1/8in sheet brass, the imitation brake magnets are cut out of 1/4in by 1/4in brass rod while the actual form of the imitation magnetic brake shoes is again easily bent to shape from 1/8in brass sheet. The suspension springs are obtained from any typewriter servicing depot and the connecting cross-bars *e* are cut from welding rod, used by all garages. Threads will have to be cut on the ends of each of these latter and 8 BA nuts fitted on *each side* to hold the whole brake assembly rigid. When the finished model tramcar starts from the stationary position, the magnetic track brake blocks, although not operative, will sway quite realistically if suspended correctly from the springs, clearing the track by about 1/16in. All such details add realism to the finished model.

Other types of brakes are used in tramway practice and one of the most common is the compressed-air system, actuated by motor-driven compressors. Quite obviously this is impracticable in a scale model tramcar but it is possible to include the storage reservoir underneath the model to add realism, if the modeller is constructing a car in which this system is used.

In an emergency, to stop a model tramcar quickly by remote control, to prevent an accident, it is possible to reverse the motors. Although purists will raise their hands in horror at such advice, it can again be emphasised that the type of motors advocated in this book are of extremely robust construction and, provided that this method is not used to excess, it is rare to find that damage to them has been caused.

6 CHOICE OF MEDIUM IN BODY CONSTRUCTION

Upon the choice of the materials used in construction of the body of the model tramcar will depend the ultimate success of the finished product. This decision, therefore, merits careful consideration. Probably no other aspect of tramcar modelling presents a comparable challenge to a scale-modeller to demonstrate his skill and attention to detail. Frequently good work and commendable patience has been bestowed on a model but the end product has not come up to expectations because it does not *look* right. In many cases this disappointing result can be traced to the initial choice of the wrong material for the bodywork.

Now whatever material is decided upon the overriding consideration is the capability of the modeller to ensure success in working in it throughout whether it is wood, aluminium, or brass. Here again the capacity of the modeller's workshop, the size and weight of the finished model and the ultimate performance of it all require prior careful thought. The materials mentioned do not, of course, exhaust the possibilities but they are the ones most commonly used as a basis. It is essential to remember that, whatever construction material is chosen, a model electric tramcar contains a large area of glass. While a number of tramcar modellers use picture-glass for glazing their models, Perspex is preferable as it is more amenable to accurate cutting and, in the case of streamlined modern tramcars with curved windows, to moulding and machining to the minutely exact fitting required. Perspex is also more robust and in the event of careless handling of a model is not so easily broken. Perspex 1/16in thick is comparatively easy to cut to shape and, if curved windows are required, these can easily be made by previously cutting a mould to the shape of the required window out of tinplate, gently heating the Perspex under the kitchen-grill to make it pliable before placing it on the tin-pattern. When the Perspex has cooled, the edges can be trimmed to shape with the carborundum grinder in an electric drill or, if necessary, filed to fit.

In considering the size and weight of a scale model, the former is easily defined, as a prototype tramcar measuring 31ft 6in overall would scale down, in 3/4in to the ft, to 2ft x 1ft x $5\frac{1}{4}$in, the 1ft in height excluding the trolley-standard and its base. The weight is not so easy to calculate. One way of calculating scale weight is to use a tonnage coefficient giving the scale equivalent of the weight of a tramcar in tons actual of the particular prototype. For instance, if the weight in pounds of the model is multiplied by the coefficient given the result is scale tons. Thus the tonnage coefficient in 3/4in to the ft scale (model tons per lb weight) is 1·82 tons so that a 3/4in scale model tramcar weighing 11lb would represent 21·02 tons (1·82 x 11 = 21·02) whereas the actual prototype tramcar weight might be only 15 tons. The discrepancy is accounted for by the fact that the weight, in modelling, of the materials used does not accord with the weights used in the prototype. This, however, can be an advantage in modelling electric tramcars, especially if they are to be used on an outdoor layout as the greater weight assists electrical conductivity between the wheels and the rails.

The material to be used in construction of a model tramcar depends considerably upon the type of prototype the modeller wishes to make. A four-wheel open-top, double-deck tramcar illustrated (plate p89) would be produced in better detail if modelled mainly in obechi wood of thickness 1/16in, 3/32in, and 1/8in respectively, with the dashes in 1/64in brass sheet. In a model of this nature, the prototype of which exemplified the acme of the coachbuilder's craftsmanship, the finely fluted corner-pillars and window frame mouldings and other equally impressive details could not be reproduced satisfactorily in any other medium, by any modeller with average tools at his disposal. The upper deck, lower deck, and platforms would have to be constructed in obechi as would the bearers supporting them. The same applies to the window-framing, side-panels, decency-boards round the upper deck, bulkheads, stair-risers and treads while the stairs may be enclosed in brass of the same 1/64in sheeting as the dashes. The use of brass for this latter renders the soldering of the relative stanchion-rails, headlights, brake-handle supports, controllers, collision-fenders and other platform equipment easier; in addition a finer finish is possible as is also a stronger model. The upper-deck protection-rails in 3/4in scale are constructed with 3/32in brass rod, the same rodding is used for the destination-box supports and the stanchion-rails of the model. The bulkheads and the sliding doors, constructed in wood, with their finely engraved panelling, provide an absorbing challenge to the skill and artistry of the modeller and it is surprising what can be achieved with three Swiss files (round, knife-edge, and flat), a razor-blade, and some fine glasspaper. The interior panelling, ceilings, floor-slats, and seats are normally constructed in wood, both on the upper and lower decks, with the exception that the seat-supports are filed out of brass-sheet where these are of the 'swing-back' type (enabling passengers to face the direction of travel) but longitudinal seats can be made entirely in wood.

Aluminium is a medium offering tempting prospects for model electric tramcar construction but it has its limitations. Certainly the bodybuilder's craftsmanship described in the foregoing model of an early open-topped double-deck tramcar could not be reproduced to the standard of perfection which would satisfy a meticulous modeller but in later tramcars, particularly those built after World War II, the fluted corner-pillars and the ornate mullions of the erstwhile gifted coachbuilders were abandoned. Machine-stamped bodywork for rolling stock had come to stay. The major difficulties of aluminium concern soldering and these are too well known amongst modellers to need any enumeration. As soldering enters considerably into model tramcar and tramway construction it is of paramount importance that the elementary principles of the art are thoroughly grasped at the outset. The ductility of aluminium is attractive but it should be borne well in mind that the subsequent difficulties, as the model progresses, will become very apparent. While metal-snips with a fine cutting edge may solve some of the problems of working in aluminium, for the ordinary amateur modeller the range of tools required for stamping out windows accurately (to leave the narrow mullions and intricate fan-lights) are of such complexity as to deter the majority, except those few who have access to equipment of this nature. The added drawback of aluminium concerns the unsatisfactory soldering aspect whereby, in consequence, liberal use has to be made of nuts and bolts or rivets and these cannot always be countersunk to a degree commensurate with a well-finished model when the painting stage is reached.

A close inspection of the well-finished models at any exhibition devoted to model rolling-stock will have impressed upon the potential tramway modeller the wide use made of 1/64in brass sheet. Together with brass rod, brass strip, brass angle, brass 'T' section, brass channel, brass tubing, brass nuts and bolts, brass screws, and a host of other essentials, it offers the amateur modeller an easily workable medium at a reasonable cost. It is also amenable to accurate drilling and easy soldering. The Liverpool Corporation 'Baby Grand' illustrated in Fig 11 is a good example of the modern type of bogie-car which would show a good result, if modelled mainly in brass. The sleek, streamlined and well-defined bodywork offers unlimited opportunities for the modeller's skill and patience. Not only the panelling which, after carefully measuring the scale, is cut out of sheet brass but the window mullions in appropriate section brass strip can be soldered on to the inside of the side-panels, both upper and lower, using the whole extent of the brass strip from lower saloon floor to upper deck roof, thus adding strength to the finished model. Smaller section brass strip can also be used to represent the beading over the seams in the panelling only. In the case of scale modelling, this is only for effect as the panels are not sectionalised as in the prototype car, for a small model does not warrant it. The collision-fenders (never referred to as 'bumpers' in tramway parlance) are also constructed in brass-strip in this type of model, afterwards being chromium-plated, a not insurmountable refinement as a glance at the classified section of any telephone directory will prove. It is also advisable to construct the lifeguards, under the platforms, in appropriate section brass strip for, although in actual practice they are always wood, experience has proved that through frequent handling the lifeguards of a model tramcar are the first parts to suffer damage, requiring constant repair, if made in wood. The lower and upper saloon floors are constructed in wood while the roof can be either wood or brass sheet. If made in wood, it is necessary to gouge out the interior while if brass is used, the central section supporting the trolley-plank would have to be an insert of wood, for insulation purposes as otherwise the trolley-pole collecting the

Page 53 (above) Tramway operating shed on a 3/4in to the ft garden scale model tramway at Glastonbury. Two ex-Glasgow full-size tramway controllers can be seen inside (one for each track). The model tramcars run straight through the shed. Good visibility is essential when operating an extensive garden layout, hence the area of glass. The model tramway depot can be seen on the left; *(below)* Birmingham Corporation Tramways stores van No 6, modelled in 3/4in to the ft scale by Mr Peter Hammond

Page 54 (above) London County Council Class E/1 tramcar No 1601, constructed to a very high standard of perfection in 3/4in to the ft scale by the late Mr F. J. Roche. The model is shown in the original state of these cars before the platforms were glazed for the protection of the crews; *(below)* scene outside the depot of Mr Richard Elliott's 3/4in to the ft garden scale model tramway at Abbey Wood, with a London United early-type tramcar No 7, London Transport No 1574 behind, and open-fronted London County Council car No 534 emerging from the depot

current would short-circuit with the body of the car which provides the return circuit. The stairways including the treads and risers can also be cut from brass sheet while the concertina-type collapsible entrance doors provide yet another example of the versatility of brass in the model field as they are easily made operative by soldering on small brass hinges on the insides. The trolley-pole is constructed from brass welding-rod obtainable from any garage. Glazing the windows, preferably as previously explained, in Perspex is, like all modelling, a job which cannot be hurried. Cutting (and in some cases machining) the windows to correct size, fixing them with Bostik clear adhesive requires delicate handling. Bostik, although a splendid and undoubtedly firm adhesive well repays for following the instructions closely. The control of its liquidity is only possible with practice and, as it dries fairly quickly and adheres firmly to whatever surface it is applied, particular care is necessary in inserting the previously cut Perspex windows into the Bostik-lined sills. Tiny blobs of Bostik visible round the edges of the windows of a completed model detract from its otherwise perfect finish. On the other hand, perfection in glazing adds enchantment to the completed tramcar.

No concern about the unsightly appearance of soldering visible in the completed interior of the brasswork of the car need worry the modeller as all is covered by 3/32in obechi wood panelling, remembering to insert the appropriate wiring for the power and lighting before completing the interior panelling. The inside of the dashes are not, of course, panelled so it is necessary to trim any surplus solder from the inside soldering of the headlamp(s).

A study of the illustrations throughout this book will show that all the models have carefully been given their correct prototype fleet numbers, as it is an unwritten law amongst all tramcar modellers that a specific car is taken and modelled. There is, of course, nothing to stop any modeller constructing a 'free-lance' model or, for that matter, modelling a foreign tramcar as many overseas systems are still operative and impressively modernised.

FIG 11 Liverpool Corporation Passenger Dept four-wheel double-deck streamlined tramcar

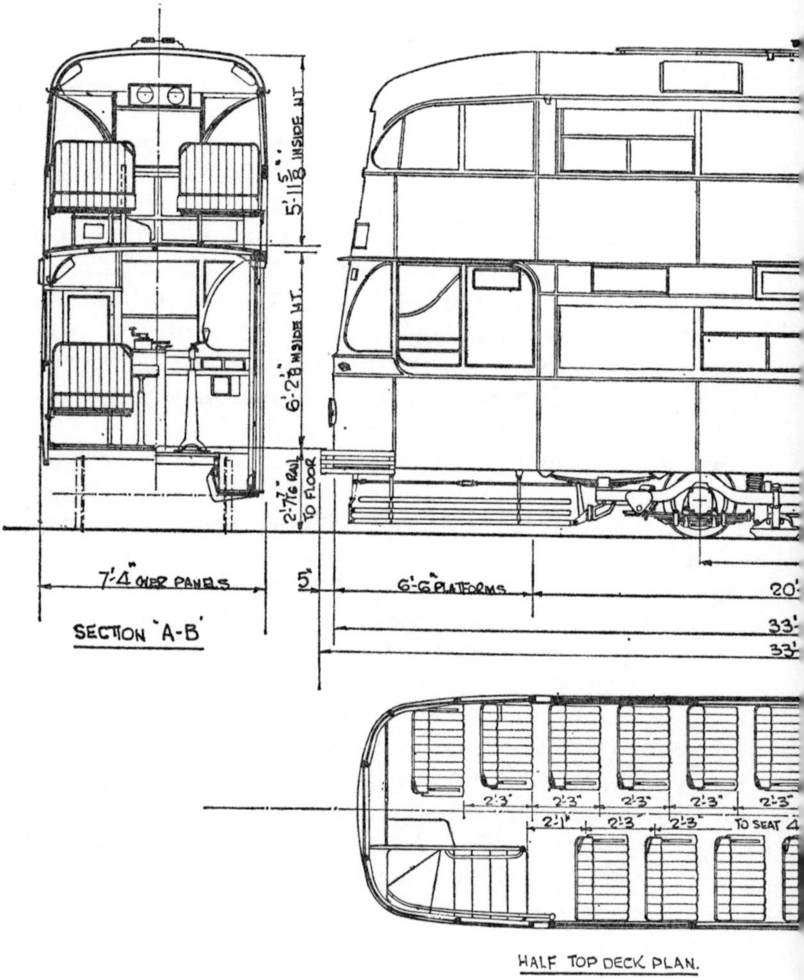

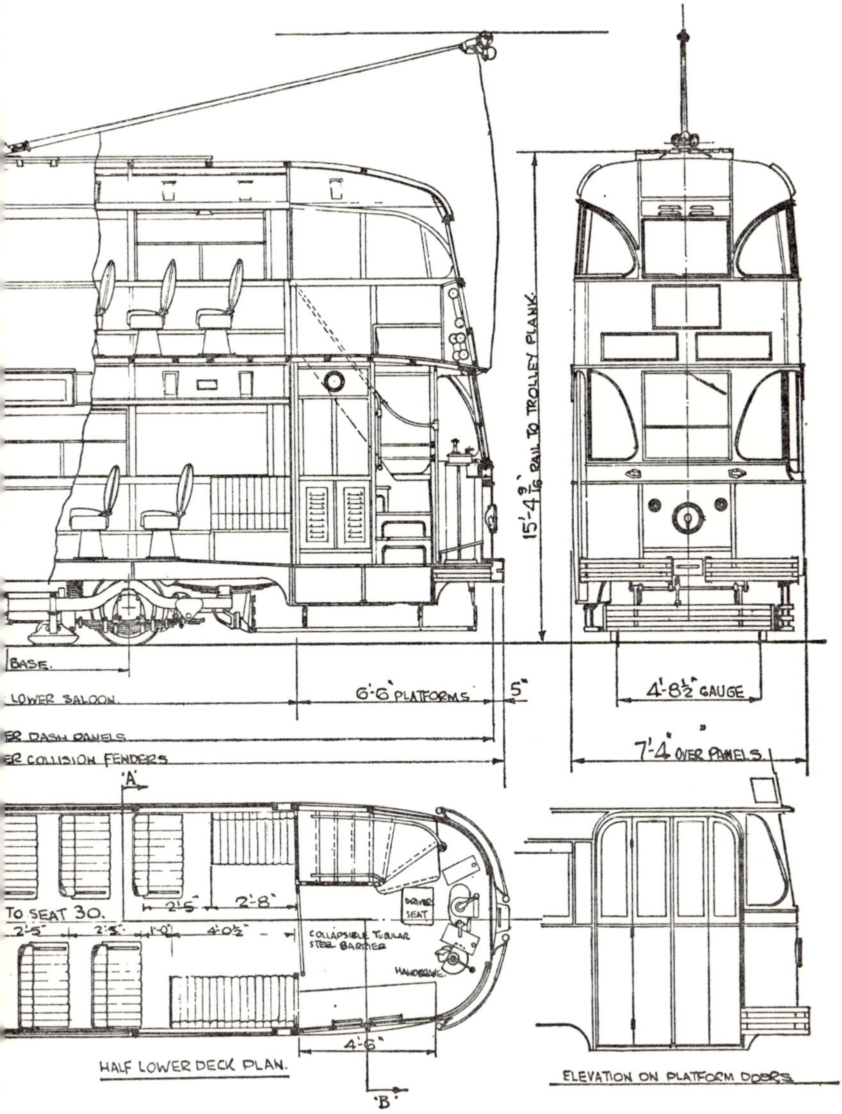
15'-4 9/16" RAIL TO TROLLEY PLANK.
BASE.
LOWER SALOON.
6'-6" PLATFORMS
5"
4'-8½" GAUGE
ER DASH PANELS
ER COLLISION FENDERS
7'-4" OVER PANELS.
'A'
TO SEAT 30.
2'-5"
2'-8"
2'-5"
2'-5"
1'-0"
4'-0½"
DRIVER SEAT
COLLAPSIBLE TUBULAR STEEL BARRIER
HANDBRAKE
4'-6"
HALF LOWER DECK PLAN.
'B'
ELEVATION ON PLATFORM DOORS

7 GENERAL CONSTRUCTION OF THE ESSENTIALS

Although, as has been stated, it is a matter of preference regarding the order of construction of a model tramcar, whether to build the car body first, or the trucks, many modellers tend to tackle the trucks first, turning their attention to the body when the truck, or in the case of a bogie-car the trucks, are completed.

In Chapter 4 the general essentials of the construction of a tramcar truck were explained. Following the completion of the truck frame the next item of paramount importance concerns the wheels and, to a large extent, the success of the model depends upon their accuracy, correct turning, and proper mounting on the axles. Inattention to any of these details will result in derailments if the profiles are incorrect or if they are not properly mounted on the axles.

Here again, a study of actual prototype tramcar practice will assist the modeller before he attempts wheel-turning. The wheel-profile of a tramcar is much smaller than that of a railway vehicle. This is due to the tramway rail groove being only $1\frac{1}{8}$in wide by $1\frac{1}{8}$in deep, except on curves where both measurements were $1\frac{1}{4}$in wide and deep, known as British Standard. The depth and width of the groove actually made street tramways possible as such shallow grooves could not affect other users of the roadways. The principal wearing portion of the wheel is the tread, and the flange runs in the rail groove serving to keep the wheels on the rails. There is one notable exception to this standard tramcar practice and this concerns American and Continental tramways where railway track is used on 'reserved' (ie private right-of-way) sections of the systems. In these cases a heavier type wheel is used similar to actual railway practice, the street paved section of the tramway having a correspondingly deeper and wider groove in consequence. Blackpool is a British example. The modeller intending to construct an outdoor 3/4in to the ft scale model tramway would be well advised to adopt the heavier type wheel profile as with the slight track distortion one occasionally gets with outdoor layouts, due to weather conditions, it is worth bearing in mind that a 1/16inch distortion of gauge will not normally derail a car, 1/8in certainly will.

In prototype tramway practice the gauge line of the rail is not the same as the gauge line of the wheel but it is slightly displaced to allow clearance when the wheels are turning. In a track-gauge of 4ft $8\frac{1}{2}$in the wheel gauge would be 4ft $8\frac{3}{16}$in.

Now many tramcar modellers are discouraged when it comes to wheel turning and mounting on axles but a little thought will overcome such fears. Some of the model engineering firms will turn and mount wheels for the modeller but this can be expensive. Whether this is undertaken professionally or whether the modeller does it himself it is of paramount importance that the standards laid down by the Tramway and Light Railway Society are adhered to as their value has been proved by many tramway modellers both on their own and on the Society's frequent exhibition tracks. The smooth precision with which the models of many members run at exhibitions is proof indeed that the standards have been justified.

Rough cast wheels, ready for turning, can be obtained by members from the Tramway & Light Railway Society or it is possible to purchase them from model engineering firms. The type of wheel depends on the prototype being modelled. Spoked driving wheels were used in some cases and disc driving wheels in others while disc pony-wheels were standard practice for maximum-traction trucks. The diameter of the wheel is understood to mean the greatest diameter of the tread. It varied in actual practice from 20 to 33in, the most usual being 30 and $31\frac{3}{4}$in diameter for driving wheels.

The enlarged section of the wheel profile in the illustration, Fig 12 overleaf is to the standard of the Tramway & Light Railway Society and it is important to notice that the shape of the flanges *must be rounded off with the back of the flanges tapered.* The tapering assists the wheel at points, frogs, and curves against the check-rails. The Society have also found it to be of paramount

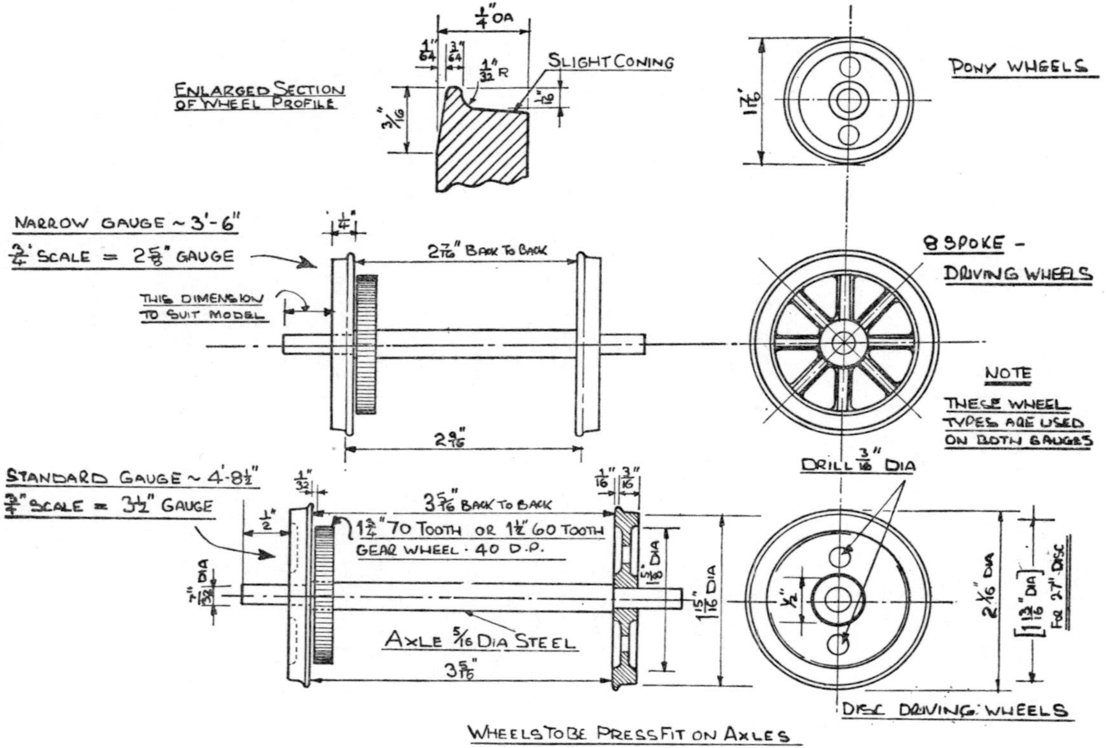

FIG 12 Wheels (Courtesy of the Tramway & Light Railway Society)

importance to insist on a standard back-to-back measurement of the wheels, when mounted on the axles. This measurement must be carefully taken from the backs of the wheels (*not* the tapering-off flange area) and should measure in $3\frac{1}{2}$in gauge $3\frac{15}{16}$in while in $2\frac{5}{8}$in gauge the measurement should be $2\frac{7}{16}$in.

It is advisable to make the axles from 5/16in diameter rod in silver steel with the wheels and gears being pressed on to the axles to ensure a tight fit. When turning the wheels some modellers first turn them separately, leaving a slight excess, before mounting them on the axles. It is then possible to press the unfinished wheels on the axle and complete the turning between the centres of the lathe resulting in perfectly exact running. Before finally pressing the finished wheels on to the axles the gear wheel should be positioned on the axle. The journals (at each end of the axles) should be machined to a smaller diameter than the remainder of the axle and about 1/16in side play allowed for freer running in the truck axleboxes.

The usual method in tramway practice is to suspend the motors on the trucks, gearing them to the car axles by means of a single-reduction gear. In other words a pinion at one end of the armature shaft drives a spur gear on the car axle, the weight of the motor thus being carried partly on the axle and partly on the framing of the truck. There are two principal methods of suspending a traction motor from the truck of a car, nose suspension and cradle suspension. In either case the motor is hung on the car axle between the car wheels, being supported at the side opposite to the axle by a suspension bar which is itself supported by springs carried on the frame of the truck. It will be seen that a flexible support for the weight of the motor is thus assured, reducing stresses on the motors and track.

It has been established that scale model tramways in 3/4in to the ft scale operate efficiently on 24 volts dc. Motors of 24 volts for this scale can be purchased either from model engineering firms or, by shopping around, from government surplus stores where they are very much cheaper. Originally the only motors obtainable were of a larger size and type than is desirable in this scale but it is now possible to purchase ideal permanent magnet motors that are $2\frac{1}{4}$in long with the outer casing $1\frac{1}{4}$in in diameter. These motors fit perfectly into both $3\frac{1}{2}$in gauge and $2\frac{5}{8}$in gauge trucks giving the clearance between the wheels for the necessary gearing. They are very powerful and stand up to all weathers and very hard running conditions. Except for occasional renewing of brushes (and it is very occasional indeed) it is unwise to attempt to dismantle these motors as they are constructed to such a high standard of watch-like precision that, without the proper delicate tools, it is impossible to assemble them again with complete satisfaction. They run for years, giving exceptionally good service, if properly used.

Some knowledge of the gear ratio of a prototype tramcar will help the modeller. The pinion on the armature shaft of the motor is geared to a spur wheel keyed to the axle. The reduction in speed from the motor shaft to the axle depends on the ratio of the number of teeth in the spur wheel and pinion. The larger the pinion gearing into a given size of spur wheel, the smaller will be the reduction of speed. The *gear ratio* of any combination is the ratio of the number of teeth in the spur wheel to the number in the pinion. The number of teeth in the pinion is always considerably smaller than the number in the spur wheel. A motor with 14 teeth in the pinion and 68 in the spur wheel would have a gear ratio of $68/14 = 4{\cdot}86$ thus the motor armature rotates 4·86 times for each revolution of the axle, or 4·86 times as fast as the axle. Differing gear ratios are used in actual tramcar practice, depending on the size of the motors and the speeds at which the cars are intended to run. Normally the spur wheel on the axle has from 4 to 5 times as many teeth as the pinion but smaller gear ratios are sometimes used if the motors are of large size and the speed of the car is required to be high. The distance between gear centres (the shaft and axle centres) is fixed for a given motor, thus the sum of the circumferences of the two pitch circles is also fixed. Any increase in the number of teeth in one gear must be accompanied by a corresponding decrease in the other as the number of teeth in the two gears must be constant. If, for example, a 15-tooth motor pinion gears with a 65-tooth spur wheel on the axle and it is to be replaced by a 14-tooth pinion, a 66-tooth spur wheel must be used. Whatever combination of gear and pinion is used, as the distance between the centres is fixed, the total number of teeth in the spur wheel and pinion must be 80. The gears would not mesh properly otherwise.

In tramcar modelling, as in all forms of modelling, some modifications in relation to the prototype have to be made as circumstances differ and it is not always possible to obtain components equivalent to those used in the prototypes. This is the case when purchasing small motors suitable for 3/4in to the ft scale model tramcars. Although in the example given the gear ratio has been shown as 4·86, as the motors used in model tramcars run at higher speeds than those of the prototype tramcars a gear ratio of 7 to 1 has been found to be more practical. It will be seen from the example given that a 7 to 1 ratio will need a 10-tooth pinion and a 70-tooth spur wheel. Again these gear wheels can be obtained from the model engineering firms and the sizes recommended by the Tramway & Light Railway Society as proving the best in practice are of 40 or 48 diametral pitch, the 48 one being of a finer pitch tooth than the 40 one.

It is most important to ensure the meshing of these gears correctly before testing them under power. In most cases a grub-screw is provided for fixing the gear wheels to the motor shaft and the axles. Although it is a very delicate operation, a hole should be drilled through the axle to take the grub-screw as, if it becomes loose, the gear wheel will rotate without gripping the axle. The same should be done with the motor pinion. While the motor shaft is only usually 1/8in diameter the drilling of so small an area can be accomplished successfully by using the tiny drills used by dentists. Most dentists have a surplus of used drills which they are willing to give away. But it is well worth the effort to save future trouble caused through gear wheels revolving without gripping. Additionally these small dentists' drills and burrs have a multitude of uses to tramcar modellers.

In Chapter 4 reference is made to PCC type cardan-shaft-drive trucks and while the basic principles of tramcar truck construction are somewhat similar to those described, the motors drive through propellor shafts and spiral bevel gears. A description of this type of transmission is given in Chapter 15 for smaller scale models.

One important factor sometimes overlooked until it is too late is the location of the driving gears in relation to the axles. They should be on *opposite ends* of the respective axles. This has to be decided *before* the motors are mounted on the axles as, otherwise, it is difficult to correct if the motors are on fixed motor-beams. Before fixing the motor-beams and the suspension-bars to the motors lay the completed truck down on the work-bench. Position one motor with its motor pinion gear wheel on your left and the other motor's pinion gear wheel on your right. It will thus be seen that the 'drive' of the truck, when power is applied, will be evenly spread as the gears will be at opposite ends of the truck. The same principle applies to bogie-trucks having one or two motors. In the case of bogie-trucks with one motor, the matching bogie-truck should have its gears at the opposite driving wheel. This sounds a simple mistake to make... but it has frequently happened, too late to rectify without considerably more effort. A bogie-truck with two motors should have the gear wheels staggered on the two axles in the same manner as a single truck car.

Mounting the motors on the axles can be accomplished by several methods. As has been stated, it is unwise to attempt to dismantle the motors. It is therefore impossible to drill the casings to take a bolted-on motor-beam. The most satisfactory method is to construct two 'collars' out of 1/4in x 1/8in brass strip, a pair of 'collars' encircling each motor with lugs to fix round the axles at each end. The lugs are made to clamp round the axles by means of a nut and bolt at each end. At the opposite end of the axle lugs the cross-beam suspension bar can also be bolted to the 'collars'. It is, of course, vitally necessary to ensure the rigidity of the motors by means of small holding screws sufficiently tightened to keep the motor in position.

When a truck is completed it will be necessary to rig up a temporary base on the work-bench and 'test-run' it for a time as it will be found that any new truck needs 'running-in' and any stiffness can be eased by the application of oil on the gear wheels at intervals and also in the axleboxes and on the motor axle-beams.

Contact with the overhead trolley-wire for the purpose of collecting the current can be made by several methods, the most common being by means of a trolley-pole or boom to which was attached a 'trolley-head' containing a trolley-wheel which revolved constantly. Some trolley-heads were known as 'fixed' trolley-heads and in this type, although the trolley-wheel revolved, the actual trolley-head did not swivel to adjust to the trolley-wires. In this case the trolley-wires were maintained directly over the tracks for obvious reasons. In America the fixed trolley-head was in general use but there were only a few examples of it in Britain, Glasgow being the largest system in this connection until they changed all their cars to bow collection.

On open-top cars a trolley standard was installed on the floor of the upper deck, centrally situated. It was about five feet high consisting of a cast-iron base bolted to the floor into which was inserted a hollow tube containing the spring and the necessary insulated cable (Fig 13). In modelling this is modified as, with only

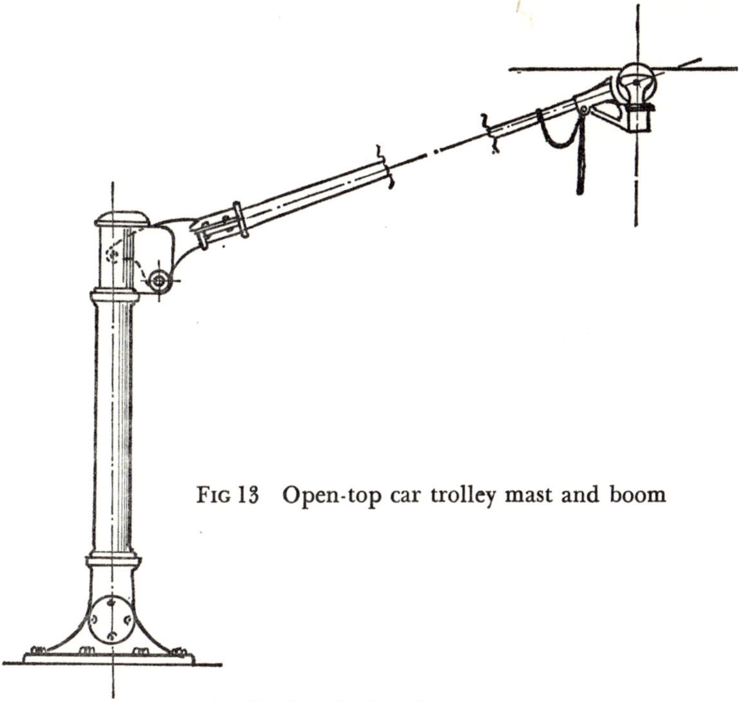

FIG 13 Open-top car trolley mast and boom

24 volts, the necessity for insulation does not arise and the whole trolley standard and boom can be made in brass. The actual standard is made from a 1/2in (outside diameter) tube cut to the scale height. A base-plate of 1/8in by 1 1/4in by 1in is next drilled at the four corners to take the holding-down screws, the corners of this being filed and rounded to shape. The hollow tube is then soldered firmly on to the base-plate, the surplus solder being trimmed off neatly afterwards. The trolley standard is then screwed down to the trolley-plank on the upper-deck floor. The swivelling 'collar' allowing the trolley-boom to swing on the trolley standard is made from a slightly larger hollow tube (about one-quarter of the height of the standard) which will fit over the top of the standard. A section of 1/8in brass rod, the length of the

trolley-boom can be obtained from any garage (where it is used as welding rod) and this is soldered on to the pivot of the swivelling 'collar'. An internal spring is tensioned inside the 'collar' to allow the trolley to swing upwards against the overhead trolley-wire. The base of the spring is anchored to the trolley-plank on the floor of the upper deck of the car.

If the model is a single-deck car or a double-deck covered top car, the trolley standards are not in the form of the column described but consist of a base only. (Fig 14.) The trolley-arm *e* is

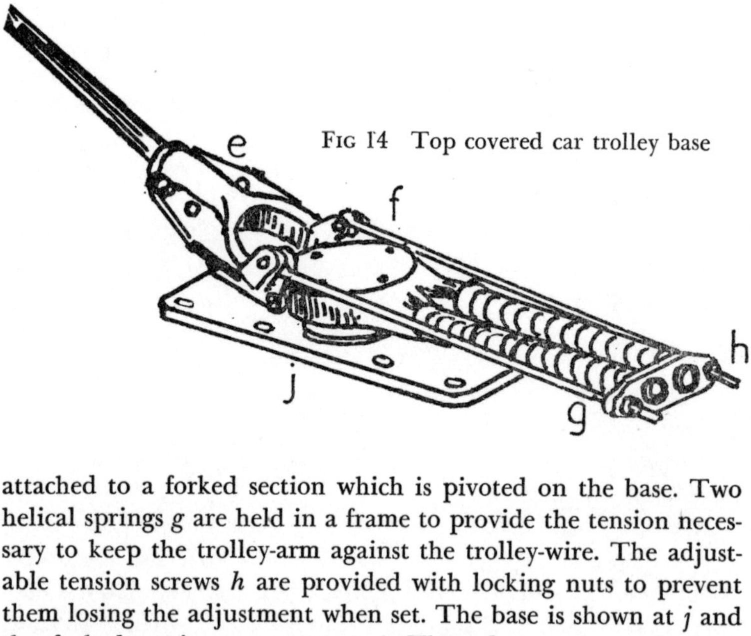

FIG 14 Top covered car trolley base

attached to a forked section which is pivoted on the base. Two helical springs *g* are held in a frame to provide the tension necessary to keep the trolley-arm against the trolley-wire. The adjustable tension screws *h* are provided with locking nuts to prevent them losing the adjustment when set. The base is shown at *j* and the forked section supports at *f*. These latter are made from 1/16in brass rod with the ends threaded to take the locking nuts.

In both of the foregoing types of trolley-arms a swivelling trolley-head (Fig 15) is provided. All trolley-heads by law had to be made detachable so they were mounted on a 'sleeve' which fitted over the actual trolley-arm. The purpose of this was that, in the event of a dewirement and the trolley arm-flying up if it struck a suspension wire, the trolley-head would detach thus preventing entanglement and bringing the trolley-wires down. To prevent a trolley-head so detached from falling into the street, a small safety-

FIG 15 Swivel trolley-head

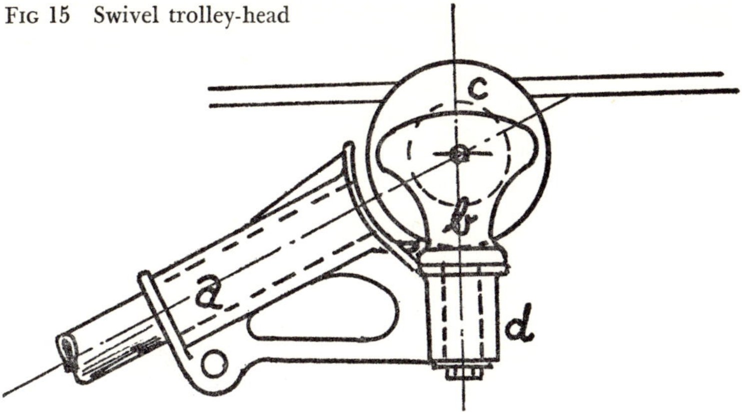

rope was affixed to it and attached to the trolley arm below the extent of the 'sleeve'. The 'sleeve' *a* is made from a short section of brass tube and the revolving globe *b* is mounted on the support at *d* relying on a lock-nut to keep it in position and allow it to swivel. The trolley-wheel itself revolves inside the globe being mounted on a spindle. Trolley-wheels in prototype practice are always made from phosphor bronze and the modeller will do well to make his trolley-wheels from the same material. It is best machined in a lathe and, in 3/4in to the ft scale 1/4in or 5/16in phosphor bronze rod should be used as brass wears down much too quickly when used for trolley-wheels.

The bow-collectors used on some tramways in England in later days and also on European tramway systems had some advantages over the trolley wheel as they cannot run off the trolley-wire. Glasgow and some routes in Birmingham were examples of the use of bow-collectors. A standard type of bow-collector is shown in Fig 16 and it is easily made in modelling by the use of 1/16in brass rod *k* with the sliding contact at the top being made from 1/8in brass rod. As the bow slants back from the vertical in practice the necessary springs *j* shown have to be tensioned accordingly to adjust to the direction of travel. One advantage of bow-collectors (and pantographs) is that they do not have to be reversed at termini. Bow-collectors are made self-reversing at termini by the simple expedient of the trolley-wire being suspended slightly higher so that the bow-collector stands nearly upright as the car arrives at the terminus. When it is reversed to commence the return journey

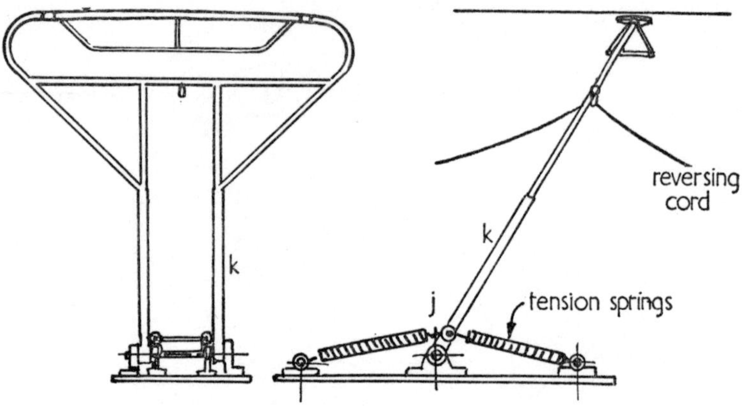

FIG 16 Bow collector

the bow automatically lifts the trolley-wire slightly, reversing itself.

A third method of current collection used liberally on European tramways and only on a few British tramways (Sunderland, Swansea and Mumbles, and some isolated instances) is the *pantograph,* which is a flexibly-jointed parallelogram shown in Fig 17. Again, this is easily made on a base of 1/4in section brass angle *z* which is mounted on the roof and drilled to take the 1/16in brass rod

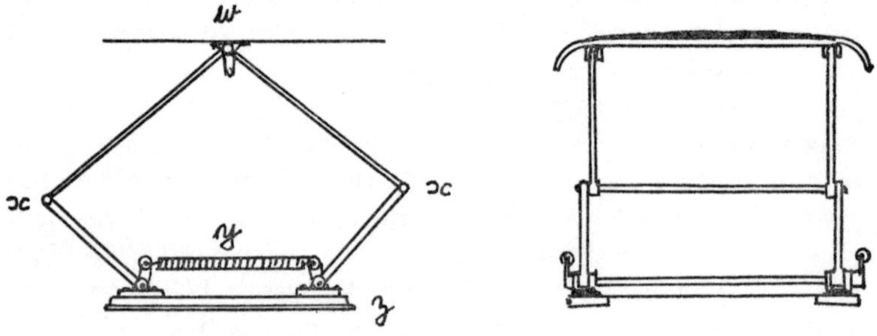

FIG 17 Pantograph

from which the pantograph is constructed. In modelling, the flexible joints of the pantograph at the corners are made by turning the 1/16 brass rod, with a pair of longnose-pliers to form an 'O' and inserting 8 BA nuts and bolts for holding purposes, remembering to allow play for flexibility and to screw on a lock-nut in each case, at *x*, the base and also at the top where the top-plate is positioned. The top-plate is made from 3/16in brass strip 3/4in wide with the edges chamfered down all round and the ends tapered to 3/8in at each end. Also at each extremity of the top-plate the ends should be slightly bent for the extent of 1/8 of the area of this top-plate, to ensure that it is guided to adjacent spans of the overhead and glides smoothly into position at points and crossings. The top-plate *w* is hinged to facilitate its easier adjustment to inequalities of the overhead trolley wire.

In the three methods of current collection described the base of the collectors will, of course, need to be insulated from the body of the car. If the car has a timber roof this is no problem but in the case of a brass or aluminium roofed model a strip of insulating material will have to be inserted. The main feeder-cable from the current collector will be taken down through the car to a distribution panel underneath one of the platforms from where the traction and lighting circuits can be fed.

8 TWO METHODS OF BODY CONSTRUCTION

Encouraged by the completion of the model tramcar truck, or in the case of a bogie-car, trucks, the modeller will be eager to make a start on the actual tramcar body. It will be noticed that, throughout this book, continual reference has been made to how tramcars and tramways were constructed in actual practice and this is very sound advice. The tramway engineers, tramcar body-builders, and ancilliary workers all kept to established basic principles which had been proved and it is unwise for a modeller to innovate as this course of action leads to an unsatisfactory model and disappointment.

So, again, it is wise to profit by the experience of past tramcar body-builders and examine their methods before embarking on body building. One of the most common types of electric tramcar used on British tramways was the four-wheel double-deck open-top car, Fig 18. The drawing illustrates a typical balcony type body built by the Brush Electrical Engineering Company Limited of Loughborough in 1904 although other notable builders constructed the same type of car. The side elevation and plan showing the seating together with the vertical cross-section show the constructional details. The seating accommodation in this car was 56 with 22 in the lower saloon and 34 on the upper deck. The seats on the lower deck are longitudinal while those on the upper deck are cross-seats with swing backs which were reversed at termini, with the exception of the immovable seats over the canopies. It will also be noticed that the upper deck seats on each side of the trolley-base were single-seats to facilitate easy passage of passengers round the trolley-standard. In later years some of these open-top cars were provided with a roof, for the protection of passengers against

Page 71 (above) Glasgow Corporation four-wheel tramcar No 38, constructed in 3/4in to the ft scale by the late Mr F. J. Roche. The decorative fanlights in both the lower and upper saloons, the meticulous lining and lettering, and the perfect crest on the waist-panel represent the acme of the model-maker's art; *(left)* Metropolitan Electric Tramways 'Feltham' type tramcar No 359, constructed by the author and his son, shown on the garden scale model tramway at Glastonbury. The tramway extends for 210ft, half of it double track

Page 72 (above) Mr C. B. Carlyle's 1in to the ft scale model of a Dublin United Tramways car running on his garden layout at Burnham-on-Sea. As the gauge of the Dublin tramways was 5ft 3in the model runs on 5 1/4in gauge track; *(below)* two 'Feltham' type tramcars passing each other at night on the Glastonbury model tramway

inclement weather, and in some cases glazed balconies were constructed over the canopies for the same purpose.

But in explaining the general principles of early tramcar body building the methods used in this well-known type of car will illustrate the modeller's requirements. It will be seen that single sliding doors were provided at the ends of the car to give access to the interior. Stairways gave access to the upper deck at both ends of the car. The upper deck was surrounded by a metal screen round the canopy and either a metal or wooden 'decency-board' at the sides of the upper deck (so called 'decency boards' as they were installed following complaints from Victorians who had glimpsed the disturbing sight of ladies' ankles disclosed on windy days as the tramcars sped down the streets). The motorman stood immediately behind the metal dash, also behind which was the controller and handbrake. The total length of this type of car was 28ft over the collision fenders.

Except for the metal used in the dashes, the upper deck metal screens and the stairways, these cars were constructed in wood and were fine examples of the coachbuilder's art. It is to wood, therefore, that the modeller must turn for his medium of construction of his scale model. The most common types of wood, obtainable from all model shops, which will be found most useful are hard woods *which have been seasoned*; obeche is good but mahogany, beech, or resin bonded plywood is preferable. These can be bought in thicknesses from 1/16in to 1/4in. Soft woods such as balsa are quite unsuitable for tramcar modelling.

Before starting to build the car body it is absolutely essential to purchase a scale drawing in exactly the same way as a truck or bogie drawing must be obtained, to guide every stage of the progress. These drawings are obtainable from the Tramway & Light Railway Society or Mr Eric Thornton.

Very great care must be taken in constructing the body underframe since if it is not measured most carefully and meticulous attention paid to lining it up properly, the whole construction of the car body will be affected throughout. The underframe may be made from mahogany or similar wood in strips of 1/4in x 1/4in section. Do not rely entirely on glueing the frame but reinforce with modellers pins to strengthen. In making this frame bear well in mind that the rocker-panels and the bulkheads have to be attached *outside* of this frame so it is usual to allow about 1/16in

FIG 18 Typical balcony type double-deck four-wheel car by Brush Electrical Engineering Co, Loughborough, 1904

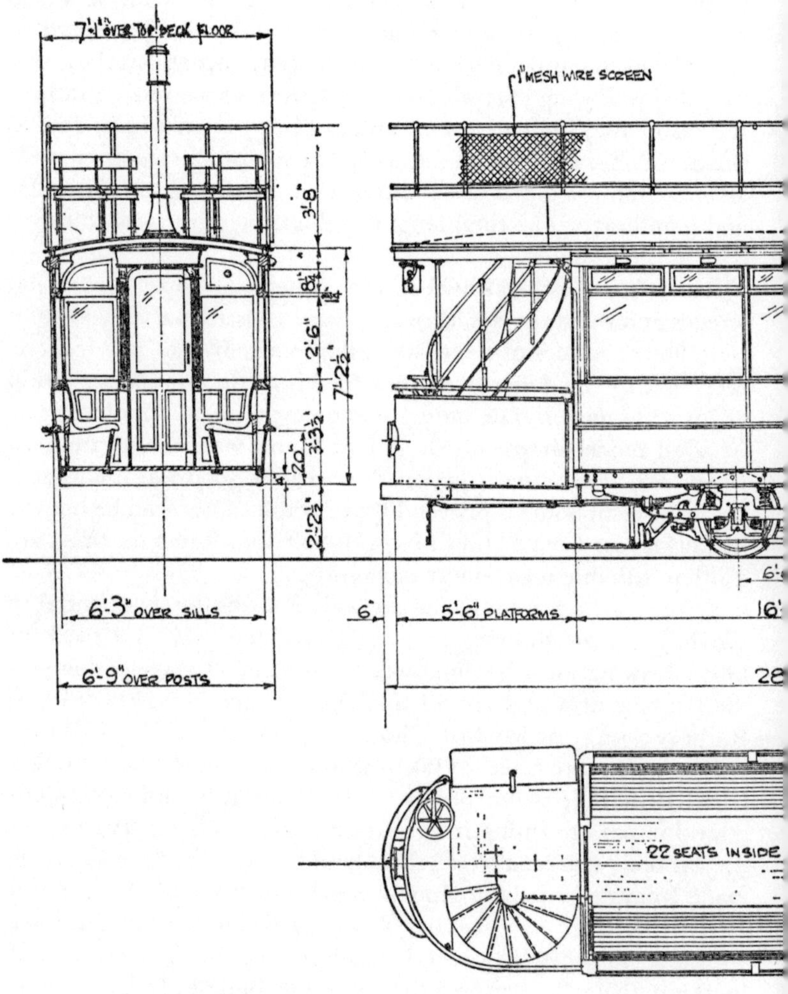

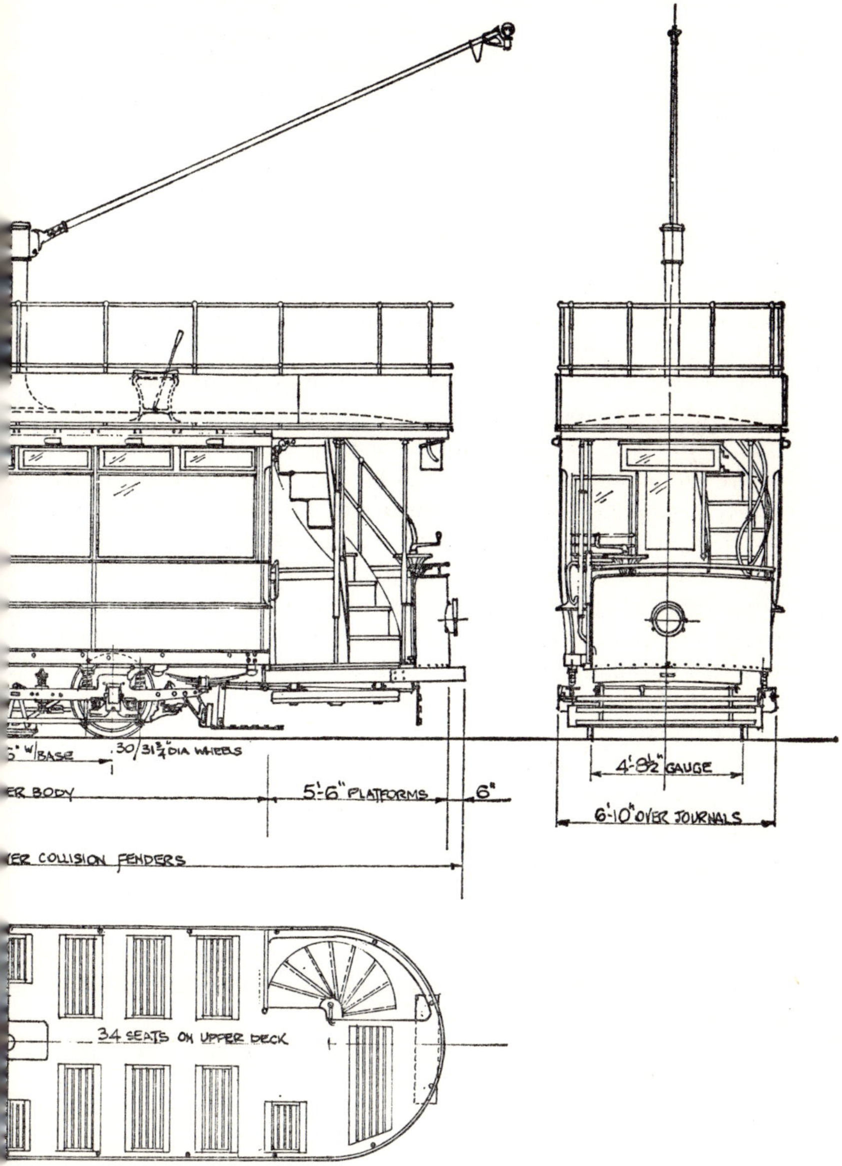

W/BASE
30/31¾" DIA WHEELS
ER BODY
5'-6" PLATFORMS
6"
ER COLLISION FENDERS
4'-8½" GAUGE
6'-10" OVER JOURNALS
34 SEATS ON UPPER DECK

in one's measurements for this. The platform bearers which support the platform are attached to this main body underframe and, here again, do not rely entirely on glue but strengthen with pins as these platform bearers have to take the collision fenders at their extreme ends. The floor is next cut from 1/16in sheet wood and glued and pinned to the underframe. The collision fenders may be made from 1/16in x 5/16in brass strip bent to shape, a 1/8in top strip being afterwards cut and soldered on to the top of each.

This process can be awkward and requires patience and the assembly will afterwards require grinding to a perfect finish with an emery wheel in the drill but it well repays the care taken as the exactness of the collision fenders adds much to the appearance of the finished model.

When this section of the model is finished the next stage is to mark out, with a pencil, the site of the seat-panels, the floor position of the stairs on the platform, the site of the controllers and where the brake-stock holes are to be drilled. If the modeller has not remembered to leave apertures for the four wheel-arches in the floor, it will be well to cut these out now before the progress of the side-panels makes this impossible. The remainder of the floor which has not been so marked out can now be covered with wearing-strips or slats normally 1/16in x 1/8in wide which should be glued into place with a space of 1/32in between each, longitudinally in the interior of the car and in a transverse direction on the platforms.

There are two methods of proceeding with the next stage. Some modellers cut and shape the corner posts first, completing the delicate 'fluting' on the outside of the posts and the 'tumble-to' of the rocker panels in this shaping. Other modellers prefer to make the entire bulkheads, complete with corner-panels in one operation. It is a matter of preference.

A number of modellers who complete the bulkhead first start by placing the shaped corner-posts in a previously carefully prepared jig and building up the framework, the door posts, lintels, and window frames, glueing and pinning as they proceed. A vertical cross-section plan shown on all scale drawings of tramcars will indicate (in the case of the four-wheel open-top Brush car we are considering) that the bulkhead under the stairway is *double* to receive the sliding door, the opposite side bulkheads being single-panel as normal. The inner sides of this double bulkhead door should receive their coats of paint and varnish before being glued into place as, afterwards, they are inaccessible. A jig should also be made to construct the doors as these have to be cut from 1/8in wood entailing delicate work in cutting out the window and also the panel-moulding. In making the doors the modeller will have cause to be thankful that he has chosen hardwood and, even so, he should not be discouraged by having to make several trial attempts both at the bulkheads and the doors, if he is new to this type of modelling. When these prototype tramcars were constructed earlier this century, the coachbuilders and coach trimmers took great pride in their art as the days of stamping all vehicles out of a piece of sheet steel had not arrived. The great care taken with the panelling and scrollwork by these master-craftsmen took time as will the dedication of the true modeller whose reward will be in admiration of the finished model, at exhibitions and elsewhere.

At this stage it is advisable to glaze the bulkheads for, as one proceeds with construction some sections of the model become less accessible. Here again there are two schools of thought. Picture glass is normally used and for the modeller accomplished in glass-cutting there is no problem. Although discouraged by some modellers, Perspex of 1/16in thickness has been found to be extremely satisfactory. It is easily workable, needing only marking out and scribing deeply *on both sides* to cut into small sections. The disadvantage of picture glass is that it breaks easily and, although modellers are very careful both at exhibitions and on their own tracks, accidents do happen and a heavy bump with the collision fenders of two models sometimes results in windows being broken or cracked. Glass windows thus broken are not easily replaceable. The disadvantage of Perspex is that it does scratch but this can easily be avoided by wiping any excess paint off with a good paint remover. Some modellers own models seventeen or more years old with not a single scratch on the Perspex. On balance, considering the handling during the life of a model, Perspex will be found to be more advantageous in the long run. A *very thin* film of Bostik is necessary to hold the Perspex windows in place. A 1/4in brass tube should be fixed against the inside of the double bulkhead doors to take the electric wiring of the car.

When the bulkheads are firmly glued into position, the doors can be inserted from the top (remembering to paint them beforehand) and they should operate easily by sliding open and shut.

With the bulkheads in place, the framing of the sides is commenced. The waist-bearer should be carefully cut to shape recessed to take the panelling and with an 1/8in protrusion on the outside of the car bevelled to the degree shown in the plan. The next longitudinal is the window-sill and this is cut to length. The car being made is a three-windowed car so the four verticals are next cut remembering that these extend from floor to ceiling. It is advisable to mitre the intersections of the framing for strength before glueing them into place. The remaining two longitudinals required are for the fanlight-framing and the top of the entire side-framing, which are similarly mitred and glued into position. It is better to leave several hours between these various operations to allow time for the glue to set. When the framing is quite rigid the rocker-panel (the lower longitudinal panel running the length of the body, which has a slight 'tumble-to' angle from the waist-bearer down to the sole-bar) is now cut, cleaned with glass-paper, and glued to the bearers. The waist-panel is treated likewise, both should fit exactly into the prepared recessed bearers. At this stage, it is wise to glaze the windows if one is using Perspex, as added strength is thus given to the whole side of the car before the sides are glued into position, with the bulkheads. The use of weights or clamps is recommended throughout the whole of the glueing operations to hold the various sections in position until the glue sets. The fanlights can be made from small 1/16in Bristol board and glazed with Perspex. It is worth keeping excess bits of Perspex which may have accumulated in the previous cutting for these smaller fanlights and destination-box windows. The necessary beading, round the windows and elsewhere, can be reproduced by using the same 1/16in Bristol board, which comes in useful for many embellishments.

The body of the lower deck of the car now completed, it is advisable to check whether all painting and varnishing of the interior has been satisfactorily completed as, from the next stage it will become impossible to rectify any omissions in this sphere.

Before proceeding any further, the longitudinal seats now have to be cut from 1/8in hardwood sheet. Both the backs of the seats and the actual seats are, as will be seen from the illustration, slightly curved. Before fixing the seats, it is necessary to glue longitudinal bearers into position to support them. Again, the actual seats and backs should be scribed to represent the lath effect in the prototype tramcars, before they are glued into position. Seats were normally varnished with a light coat of varnish and here again, it is necessary to emphasise that this should be done before they are fixed because interior access is impossible later. The floor should be painted with a dark grey undercoat only.

The car ceiling now has to be measured and cut to shape and the 1/8in obeche sheet can be used again for this purpose. Here again, a look at an actual prototype tramcar will help the modeller as car interior ceilings gave much scope to the designers. Some were just 2in slats, recessed at the ends, being alternately varnished dark-oak and light-oak giving a very pleasing effect. In later years these slats were painted white on some systems to accentuate the interior lighting. On other systems ornate designs of diamond-shaped moulding were carved, being varnished in different shades to the actual ceilings. The only guide here is to visit the Tramway Museum at Crich, near Matlock, and inspect the wide variety of interiors of the museum cars, making notes accordingly. Slats can be scribed on the obeche sheet. Mouldings would have to be cut out and glued into position. The necessary holes for inserting the lighting fittings should be bored in the ceiling and strap-hanging rails inserted. Again, it is necessary to labour the point, the ceiling has to be painted and/or varnished, whichever is decided on, at this stage.

The ceiling completed, the necessary 24-volt interior lighting bulb-holders should be fixed into the holes in the obeche drilled for this purpose, the ceiling glued into position, and the bearers for the car-roof constructed. These consist of four main bearers positioned over the bulkheads and over the vertical window pillars, consisting of 1/4in square lengths recessed to take the wiring for the traction and lighting current. With the bearers in place glued to the exposed side of the car ceiling the lights are wired *in parallel* by soldering light-gauge bell-wire to the two contacts of the bulb-holders, ie one length of bell-wire is soldered throughout to the screw-type brass holders and another length of bell-wire is soldered throughout to the small terminal contact of the bulb-holders (a radio-type soldering iron is needed for this latter delicate operation).

The floor of the top deck can now be cut out of 1/8in obeche in one entire piece, the canopy being rounded as indicated in the plan. The access for the stairs from the lower platforms must also be cut out and then the whole floor can be marked out like the lower saloon floor carefully measuring the seat positions which, in this case are the transverse type. The wearing-slats extend practically the whole length of the floor longitudinally but the wearing-slats between the seats are transverse. One item sometimes forgotten is the 'trolley-plank' which can be made from 1/4in hardwood and extends for 1/8in more than the size of the trolley-base. It is rounded at the edges (to prevent injury to passengers) and its purpose is to give added strength to the trolley standard. After glueing it directly central and drilling it to take the trolley-standard, the slats are finished flush to it. Each side of the trolley-base are *single* seats so that the *longitudinal* slats extend each side of the trolley-base accordingly.

Extra bell-wire should have been left to wire up the destination boxes if they are either on the front guard rails, or under the canopy (depending on which system the car is representing) and an *entirely independent* and differently coloured wire, sufficient to take 5 amps, inserted through a previously prepared hole in the middle of the trolley-plank. This is for the main feed of the car. The whole of the wiring, positive and negative lighting feeders and the main feed should now be threaded through the 1/4in tubing which was previously inset into the one bulkhead behind the door pillar.

The canopy decency-panels can now be made out of 1/16in brass sheet and bent round to shape. Some modellers also make the side decency panels out of brass sheet but they can also be made from 1/16in wood sheet if this is desired. The vertical stanchions are next cut from 1/16in brass rod and, after drilling the necessary holes round the edges of the floor can be inserted tightly into these. Two guard rails which run round the whole length of the upper deck are soldered into position by dropping a tiny blob of solder on to the top of the stanchions and quickly pressing the guard rail on to it before it sets. With a little practice it is possible to drop another blob of solder on the positioned guard rails to represent the support fitting used in actual practice.

Stairs are regarded with awe by new tramcar modellers because they are spiral. By referring to actual prototype practice again, this aspect can be simplified by experience. The sides of the stairs should be cut out *first* from 1/16in sheet brass, carefully measuring their spiral twist and shape from the plan beforehand and, if necessary, making a cardboard pattern. The bottom, middle and top risers should be cut from the same 1/16in sheet brass and firmly soldered into place as a first stage. From the rigid foundation thus made the treads and risers of the remaining stairs can be fixed after first cutting them into the necessary shapes and ensuring that they fit at right angles in relation to each. In sticking the wood treads and risers to brass remember it is necessary to squeeze a film of Bostik or Araldite on each separately. Remember that the stairtreads slightly overlap the risers throughout. The stairs should be painted the same dark grey undercoat as the rest of the floors of the car before they are finally fixed into place as they will be difficult to paint afterwards. Before fixing the stairs into place make sure the outsides of the bulkheads are painted in the correct final livery. Also before fixing the stairways the necessary stair handrails have to be constructed from 1/16in brass rod in exactly the same manner as the upper-deck guard rails. By making the sides of the stairs in brass strip the attachment of the handrail stanchions by soldering is simplified.

The dashes can now be cut from 1/16in brass sheet after first scribing and marking out in accordance with the plan. It will be noticed that they do not fit flush to the car body but a slight gap is left here, the beading on the top of the dashes and the rivets round the floor edges securing them. Carefully mark out the headlamp position and then drill through its centre, gradually opening the hole by means of a small modellers 1/4in emery grinder to the circular shape necessary. The actual headlamp can be made from the same 1/16in brass strip to form a circular collar, soldering it into place and making the headlamp rim from bell-wire soldered on to the completed headlamp. It is then possible to cut a circular piece of Perspex for the headlamp glass. The raised brass bars connecting with the upright stanchions at the platform entrances are made from the same 1/16in brass rod as are the stanchions and hand-rails, being soldered accordingly. Again, before fixing the dashes into place, remember to paint the inside of them either light grey or red oxide, according to the system being modelled.

Returning to the upper deck a length of 1/8in-gauge brass mesh will have to be obtained and cut to size to solder on to the stanchions above the decency boards and the top of the guard rails. Some systems, notably Manchester, Birmingham, the Metropolitan system in London, and others had ornate iron scroll work instead of brass mesh and if the modeller is contemplating this type of embellishment it is necessary to construct it in thin brass strip, first making a jig as this scroll work was in standardised sections. With the aid of a jig the ornate patterns can easily be soldered together. It needs considerable patience but the final result is very impressive.

The garden-type seats on the upper deck will take some time to construct. The basic principle is quite straightforward but 18 of them in this particular model are reversible, as their backs slide over for the passenger to face the direction of travel. The seat-legs are made from 1/16in brass strip bent to shape with the cross-sections near the floor soldered on to them as supports, after first having been drilled to take the reversible back supports. For the 18 seats 36 one-piece legs and back support uprights have to be made. The back supports, first bent to shape, can have 1/16in hardwood backrests glued to them with Bostik and the same hardwood can form the actual seat itself. As the floor of the upper deck is slightly convex (to allow rain water to disperse) care is necessary in making the seat legs as obviously they will have to be slightly higher one side than the other in order that the seat itself may be exactly level.

All handrails, brake handles and stocks, brake stock brackets and (if equipped) slipper-brake wheels can be made from the correct size brass rod. The controllers, which are dummy, can be made from 1/16in sheet brass, shaped and soldered accordingly. The controller handles can be made from brass strip and filed and bent to shape. They should be mounted on a spindle to allow them to turn in the controller tops and these latter can also be cut from brass sheet. The controller is supported against the dash by a soldered bracket. It will be necessary to make a small cone for the back of the headlamp, drilling this to take the bulb wiring and the bulb. It is also necessary to drill through the platform floor for the headlamp wiring to be threaded through to the main feeder underneath the platform. The electrical circuits for powering and lighting the car are explained in Chapter 14.

The second method of model tramcar construction concerns the use of brass throughout for the 'shell' of the model, with the exception of the floors and roof. This method is to be preferred for outdoor model tramways which operate in all weathers as the models are not affected by outdoor conditions. The method of construction is on the same principles as making a model tramcar mainly of wood except that instead of glue and pinning, solder is used for all aspects of construction.

To take the example of a double-deck covered-top tramcar of a standard design, the floor is constructed in exactly the same manner as already explained. As, in later years, many of the more modern types dispensed with rocker panels as distinct from the waist panels, the whole of the lower side of the car can be cut from 1/16in brass sheet bent to form the lower bulkheads in one operation. The whole of the upper deck decency panels, including the rounded sections over the canopies, can likewise be cut from one section of sheet brass, although it is necessary to sectionalise the sides and ends for easier modelling. The brass lower panels are then laid on the work-bench and the window pillars in the form of brass strip are carefully measured and cut into correct lengths for both lower and upper saloons (as the car is of the covered-top type) in one section for both decks. After scribing the position of the pillars on both lower and upper side-panels, they are then soldered on the inside areas. The ceiling and upper-deck flooring are cut from *one piece* of 3/8in hardwood. This is necessary to give the car strength as brass is a heavier material. A number of 'L' shaped small angle-brackets are then made from 1/4in x 1/8in brass strip and drilled at the foot of the 'L'. These are soldered inside the prepared side sections to attach the side-sections to the floor and to the upper floor. The rounded balcony is treated likewise with these 'L' shaped brackets but before it is positioned the window pillars are soldered in place.

The roof of the car is shaped from a section of wood 5/8in thick, rounded at the ends and curved on top in accordance with the plan. A number of modellers have adopted a technique for making curved sections of roofs by fixing curved sections of small-gauge brass mesh, cut and shaped to size, to the roof and side-bearers of the model, and moulding the shaped roof round this 'reinforcement' with 'Polyfilla'. This is on the principle of reinforced concrete laying. Paper of a thickness which can easily be bent to curved shape is used to hold the underside of the 'Polyfilla' in place and a mix of the correct consistency is easily pliable and moulded over the brass mesh to the domed shape of the roof of the model. One important point to remember is to make the brass mesh reinforcement frame of *smaller* dimensions than the outer shell of the roof required. When the 'Polyfilla' is dry it is smoothed down with 'wet-or-dry' fine glasspaper used wet. If the smoothing down is done with care and a paint undercoat applied, the final painted finish will resemble the painted steel of the prototype tramcar being modelled. The lower bulkhead panels being of brass easily take soldered-on handrails while the upper bulkhead window sills and the doors can be of 1/16in obeche as previously described. The bulkhead pillars will easily glue to the floors and ceilings and the glazing in Perspex is exactly as described for the previous model.

The ceiling being of one section will have to be drilled for the lighting bulbs which are wired up as in the previous car. A false floor in 1/16in obeche will have to be provided to hide the wiring and the seats affixed to this, on the upper deck.

The upper deck lighting requires care as it is necessary only *partly* to drill through the 5/8in roof to insert the necessary bulbholders to their standard screw diameter but it will be necessary to drill a smaller hole straight through, at the end for the purpose of making the contact slightly protrude outside the roof. For wiring purposes a 'V' shaped groove is cut along the whole length of the roof, in line with the bulbs, both inside and outside the roof. Into this recess is inserted fine gauge uninsulated bell wire both inside and outside, the wooden roof acting as an insulator. The 'V' shaped recess now containing the lighting circuit is filled in with plastic wood, after the bulbs have been soldered to the wiring. The slight protrusions on the outside of the roof are concealed with ventilators easily carved from 1/16in wood sheet. As the interior of the car is panelled, all wiring, both the main feed and the lighting circuit, is concealed.

If one is worried about a bulb burning out, it is worth recording that a number of tramway modellers have never had a bulb burn out in nearly 20 years life of their models. Quite true, they have treated their models properly and have never overloaded the circuits.

The same principle for screening and vestibuling the platforms can be adopted by using the correct section brass strip.

Now a model built in this medium will be much heavier than one built entirely in wood. If it is intended to run an outdoor model tramway this is a great advantage as heavier cars, on an outdoor track, provide better conductivity and give less trouble in running.

Common to both wooden-built and brass-built cars are the lifeguards. Prototype lifeguards were made of wood attached to steel frames. In the case of model tramcars it is better to make the entire lifeguards, both gate and tray, of correct section brass strip, soldering firmly together because this area of the model, with handling, is attended by the most breakages if made in wood. The sections of the slats differed with different systems so reference to the plan is advised. Some lifeguards were constructed in steel mesh (Bristol and London United, this latter in earlier cars) and the 1/8in mesh previously used for the upper-deck screens in the open-top car is ideal.

British tramcars were obliged by law to be equipped with lifeguards of an approved design able to pick up, with the minimum of injury, any pedestrian falling in front of a moving car. These lifeguards were known as the gate-and-tray trigger-type and they are quite automatic, not depending on the driver for operation but actuated solely by the obstruction on the track, in most cases an errant pedestrian.

FIG 19 Lifeguard

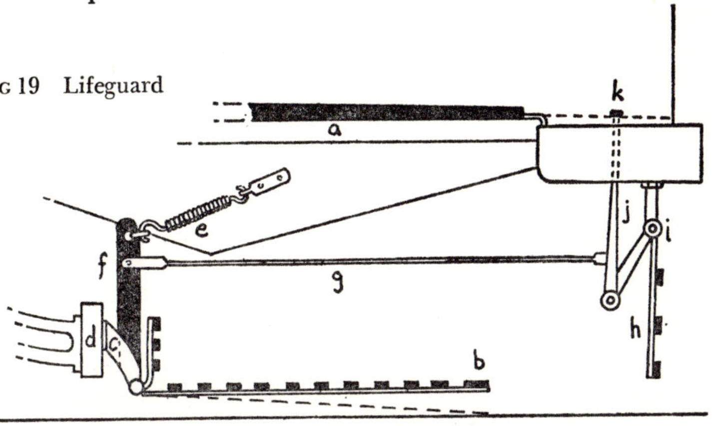

A number of modellers in 3/4in to the ft scale do not make their lifeguards operative and in smaller scales it is practically impossible to do so but as they are an integral part of the model it is essential to fit them, whether operative or otherwise. For those modellers who wish to make the lifeguards operative the illustration, Fig 19, shows the principle on which they work. They were always fixed under the platform *a* of the car and the following tray *b* was fixed a few inches above rail level pivoted to the bracket *c* which was attached to the pilot board *d* of the car truck. The helical spring *e* was attached to the car-body bearer and pulled on the lever *f* which was fixed rigidly to the lifeguard tray *b*, the spring tending to lower the falling tray and keep its front end down on the roadway. Any movement of the lever *f* is prevented unless the push-rod *g* is actuated. The gate *h* is pivoted at *i*; it is pushed back on striking any obstruction and the crank-lever *j* is rotated thus allowing the spring *e* to force the rod *g* forwards and bring the tray *b* into action by lowering it. When the obstruction has been removed, to take the weight of the tray *b* it is re-set by depressing the platform pedal *k* which rotates the lever *j* in a counter-clockwise direction putting the rod *g* into compression and in this manner raising the tray *b*. Each time the lifeguard is operated the platform pedal *k* is forced upward.

In the early days of tramcars before this type of lifeguard was invented a number of systems equipped their tramcars with a tray only. In these cases the tray was not constructed of wooden slats but was made of steel mesh which, although fixed only a few inches above rail level was not automatic in operation and relied on the weight of the obstruction to hold it down. This type of lifeguard was superseded by the gate-and-tray trigger-type and was not in use for very long.

FIG 20 Side-guard

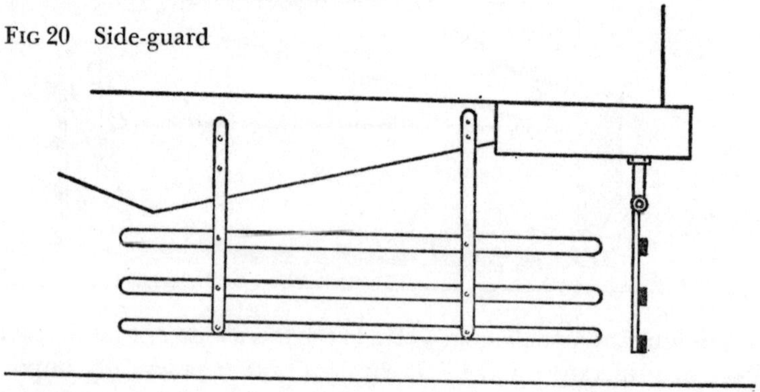

On some tramway systems, particularly in large cities, *side-guards* were provided at each side of the lifeguards (Fig 20). The type shown in the illustration was fairly standard except that in a number of cases the bottom slat was only about half the length shown. This was to obviate its fouling the tray if the side-guard was situated more towards the centre of the underneath of the platform. The side-guard on the opposite side was attached to the underside of the tramcar step if the step was of the hinged pattern which was tilted to turn up when the car was running forward. If the tramcar had a fixed step which was in a permanent position, usually no side-guard was fitted on that side of the car, or (in the case of the London 'Feltham' type cars) very much smaller side-guards were fitted as the extra long steps on these cars were also very low thus obviating the provision of side-guards.

Some systems provided side-guards between the bogies of their bogie cars which were also known, in this case, as 'dog-guards', intended to prevent roving dogs undertaking a tour of inspection underneath the car when it was stationary to the discomfiture of the dog when the car started. Although these side-guards were generally constructed in the same type wooden slats as the life-guards, a few systems had wire mesh ones, particularly so in the case of their dog-guards.

Glasgow tramways constructed some post-war II tramcars with side-guards running the entire length of the car throughout. Before attempting this, if the modeller is making a Glasgow car of this type, he should ensure that his curves are of sufficiently easy radius to enable the bogies to swivel without fouling the side-guards. This will apply to his points and crossings for a similar reason. Normal type short section side-guards are clear of the bogies and any swivel will not affect them.

Lifeguards and side-guards (where fitted) were made detachable to enable cars to be temporarily fitted with 'V' shaped snow-ploughs when required on a proportion of cars of some systems which did not retain special snow-plough cars. The movable parts of the lifeguards were detached and the snow-plough was installed underneath the platform, resting on the track and pushed by the pilot-board of the car. This can easily and effectively be reproduced by the modeller and is extremely efficient in clearing the tracks on an outdoor layout.

The destination boxes can be made from obeche and glazed with Perspex; it will be found that one single 24 volt bulb will be sufficient to illuminate the blind. In prototype practice either two or three bulbs were used but one is sufficient in a scale model tramcar.

Again common to both methods of construction some modellers *seal* the interiors of their models making them inaccessible for all time. This is not advisable as, occasionally, adjustments to interiors do become necessary when a seat becomes unglued or an interior light bulb burns out. For this purpose it is possible to make one side of the model detachable. This can easily be done during construction and there is really no reason why a precision finish should not be attained although the whole side of the car is made to be unscrewed and be opened. Ingenuity is required in all forms of model making and a challenge in this field is always worth accepting merely for the satisfaction one gets out of ultimate achievement.

Recently a number of successful model tramcars have been constructed wholly or partly in fibreglass, particularly the curved roof-sections. The first essential is to construct an *outer* mould to which wax, or a liquid release agent must be applied and allowed to dry thoroughly before starting. A small amount of hardener is applied to the resin and the fibreglass matting is cut to size, in one layer, to fit the mould. The resin is brushed well into the fibre and then more matting and resin is added until the required thickness is reached. The whole is then left to cure and, when completely dry is removed from the mould.

Now although the general principles of model tramcar construction have been described it must be remembered that the prototypes taken to exemplify tramcar modelling do not, by any means, exhaust the possibilities of what the modeller may have in mind. There is a large number of modern tramcar builders in existence abroad where these vehicles (sometimes by their modern appellation of 'rapid-transit') have been modernised, streamlined, and equipped with the very latest recent improvements. The modeller may have in mind the construction of an up-to-date European or American vehicle, possibly one of the articulated (six-axle) German or similar tramcars. From the basic information given in this book together with the necessary drawings obtainable (as a number of modellers have done) directly from the overseas builders or municipalities concerned, the modeller should experience no difficulty in constructing whatever model he has in mind.

Page 89 (above) Wolverhampton Corporation single-deck four-wheel tramcar No 69, constructed in 3/4in to the ft scale by Mr Peter Hammond and shown on his extensive garden model tramway at Luton; *(below)* Bristol Tramways double-deck open-top four-wheel tramcar No 160, constructed in 3/4in to the ft scale by the Reverend N. Jackson-Stevens. Note the 'tray' type lifeguards with which the earlier Bristol tramcars were equipped before the 'gate and tray' type became standard

Page 90 (above) Blackpool Corporation double-deck tramcar No 257 passing single-deck railcoach No 319 on the 'reserved' sleeper section of the Glastonbury garden tramway; *(below)* Mr Terry Russell's gauge 0 scale model exhibition layout

9 TRACK CONSTRUCTION AND POINTWORK

Upon the method of permanent way construction and the care and accuracy with which it is undertaken will depend the amount of trouble-free running and the ultimate pleasure obtained from a model tramway. If the original permanent way has been hastily constructed without attention to detail, constant derailments and poor running will ensue so it is far better to give time and patience to it at the start.

The section of tramway rail adopted throughout the world is known as the grooved girder rail as illustrated in Fig 21, where *a* is the flange, *b* the web, *c* the tread, *d* the check, and *e* the groove. Now this type of tramway rail is very expensive to construct as, in 3/4in to the ft and similar scales, it has to be machined out of the solid with appropriate machine-tools. It has been done by some modellers, notably those who are expert engineers having access to well-equipped machine shops. For the guidance of the amateur model engineer, therefore, it is proposed to describe a far simpler and far less costly type of construction which will achieve the same result.

Most model tramway engineers construct their tracks in brass although aluminium has lately been coming to the fore. If the track is constructed in steel, it is quite unsuitable for an outdoor tramway because of its tendency to rust with a consequent loss of conductivity. It is also necessary to clean the rails before each running session, with Thawpit or emery cloth, a tedious undertaking. This also applies to steel track of a portable layout, if it is stored indoors for any length of time. Aluminium rail is satisfactory if the resistivity is 3·7 microhm centimetres at 20°C, or less, but the most satisfactory type of rail for both indoor and outdoor

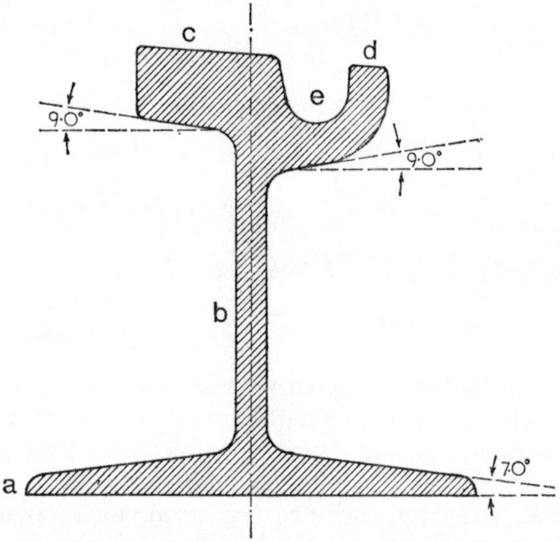

FIG 21 Cross-section of rail used by Bristol Tramways & Carriage Co Ltd. Reproduced one-third full size

use is brass 1/2in scale 'Vignoles' railway track. This type of track is one of the easiest to lay as it can be laid directly on to the sleepers. Also, due to the deep section and wide base, it is extremely strong. Provided that the sleepers are correctly spaced it does not need constant attention. It is never wise to economise on sleepers as this results in track distortion and can be a source of continual extra work on the permanent way.

Commencing with the sleepers, although they may be cut from any type of hardwood, if the layout is out of doors, oak is by far the best material. It has to be seasoned oak as the modern kiln-dried oak will be found to be unsatisfactory. The purchase of an old oak gatepost or a similar piece of seasoned oak is preferable and, after measuring and cutting a pattern of the sleepers required to the relative gauge, it is advisable to 'sub-contract' this sleeper cutting to a local joinery works, stating the quantity required in batches of a hundred depending on the length of track to be laid. Sleepers for point work, it should be remembered, are longer in section than the normal sizes cut for the single tracks.

The next stage is to procure a shallow 'drip-tray' of the type used by garages. This is then filled with creosote poured on to the sleepers to be 'pickled', taking care to cover them thoroughly, and left for a few days so that the cresote impregnates the sleepers thoroughly. It is quite useless to paint the sleepers with creosote as, on an outdoor tramway, the effect of the elements renders this protection of no value. When the sleepers are completely soaked they should be taken out of the drip-tray *singly* by means of an old pair of fire-tongs and laid out to dry, preferably on a concrete or stone pathway. When they have dried in the sun, it will be necessary to turn them over to dry out the other sides. Meanwhile, of course, the next batch of sleepers are being put into pickle in the drip-tray and so on, until a sufficient number have been so treated.

Rail sections are normally sold in three-feet lengths and, before commencing to lay rail a rail-gauge and a jig are necessary. The rail-gauge is easily made from 1/8in brass or steel bar correctly cut to fit the tread of the rail. The jig is made up from a three-foot length of timber preferably an inch in thickness and of a width appropriate to the gauge. Sections of wood of the same depth of the sleepers but *of twice their width* are then screwed into the prepared base, leaving appropriately measured gaps for the reception of the sleepers which will be inserted as the track-laying commences. It will easily be seen that this simple procedure standardises the correct spacing of the sleepers throughout. The sleepers are then inserted in the gaps of the jig and the first rail correctly aligned ready for fixing to them.

Several methods of fixing the rail to the sleepers are possible. Some model engineering firms recommend using small round-head brass screws, or a standard pin and washer. The overriding consideration, however, is to ensure that the rail is fixed permanently and firmly to the sleepers and the most satisfactory method is to drill straight through the base-flange of the rail and drive in 5/8in brass screws. The time taken will be amply rewarded by the subsequent rigidity of the track. It is only necessary to drill and drive one screw on one side of the base-flange in each sleeper (as distinct from a screw each side of each single rail) as this will be found quite sufficient and firm. One rail having been laid, the track gauge is now brought into use and the corresponding rail laid similarly. When this is completed the three feet of completed pre-fabricated rail section can now be lifted out of the jig and the process repeated to the extent of the length of track layout required. If the three-foot sections are constructed exactly as described it will be found that all possible track distortions and subsequent maintenance are completely eliminated. Furthermore no 'tie-bars' or 'tie-rods' normally used in tramway practice are necessary.

The foregoing method of construction is used for 'reserved' track where the tramway does not run on a public highway. Some tramways used this type of track on their own private right-of-way and were allowed to do so for single-deck cars only. If double-deck cars were used on a 'reserved' or private right-of-way section of tramway a 'check-rail' was required by law in some countries, the check-rail being exactly as in normal tramway practice described at the beginning of this chapter. Blackpool tramway on its 'reserved' section is laid in standard *railway* track equipped with a special type check rail throughout.

For the purpose of a model tramway and to obviate unnecessary expense as scale machined tramway track presents considerable problems, the example of Blackpool can be followed and an independent check-rail fitted to the railway-type track available from the model engineering firms . . . for none of them produce scale tramway track.

The necessity for a check-rail for scale model tramways is because of the paving at street level if the modeller intends faithfully to reproduce actual practice. The most satisfactory method of inserting a check-rail as an imitation of the prototype rail is to do so by means of three-foot lengths of 'L' shaped aluminium cut to the correct section to fit under the base-flange of the rail and trimmed level with the tread of the rail. Aluminium to these lengths is stocked by most local coach-builders and the coach-trimmers will cut and bend it to the required specifications for a quite moderate charge. It is then inserted underneath the rail previously described *before* attaching the rail to the sleepers, the same screws being driven through both the rail and the base of the 'L' section. The representation of street paved tramway is then made up by inserting a sheet of wood to rail level and painting it with a coat of flat dark grey paint. The final effect is most realistic.

The fishplates supplied for most scale track up to 1in to the foot scale are usually of a spring steel clip-on type which are inserted at the ends of the base-flange. Before fitting these it is advisable slightly to chamfer the ends of the base-flanges at each end of the rails and the fishplates will be easier to fit. In larger scales the fishplates are the same as actual practice with the accompanying nuts and bolts for fixing into the already drilled rails. Whichever types are used it is most essential to *bond* the rails for the purpose of good electrical conductivity. The bonds are cut from ordinary 18 swg copper wire and firmly soldered to the ends of the rails. The most preferable type of solder to use in all tramway modelling is five-core Multicore 40 tin/60 lead. If the ends of the rails are first cleaned with emery-cloth and tinned and the bond is also tinned, a good joint will be the result. A lot depends on good conductivity in a model tramway, especially outdoors and good bonding (and good soldering) is the basis of this. Although the modeller may be tempted to rely solely on the fishplates for conductivity, this is entirely unsatisfactory; there is no substitute for sound bonding throughout the layout. The spring steel clip-on fishplates should be coated with red-oxide paint before use as they have a tendency to rust.

All model tramways have a number of crossings, curves, branch-offs and crossovers which, with care and patience, any modeller can make, thus saving considerable expense in purchasing them ready-made (Fig 22). The simple curve *a* is easily made by first permanently fixing down the straight track already described at one end and then joining the *outer* rail of the proposed curve to the outer rail of the straight track. The rail to be curved is then carefully bent to the extent of the curve required and temporarily fixed in place by a few nails into the baseboard. The sleepers are then positioned and the drilling and fixing undertaken. The *inner* rail is then measured for gauge with the aid of the rail gauge and it will be seen that this inner rail requires a small section to be sawn off with a hacksaw to bring it level with the outer rail already laid. This inner rail is then laid with the aid of the gauge throughout the drilling and attachment to the sleepers. If the curve is laid in railway-type track only one 'check-rail' will be necessary and this is shown in *a* on the *inner* rail of the curve, which is a left-hand branch; the converse applies with regard to a right-hand branch as the check-rail has to be on the *inner* rail of the curve always.

The point *b* is a left-hand branch and, for the construction of this the same jig which has been made for the straight sections can also be used remembering that the longer sleepers used for

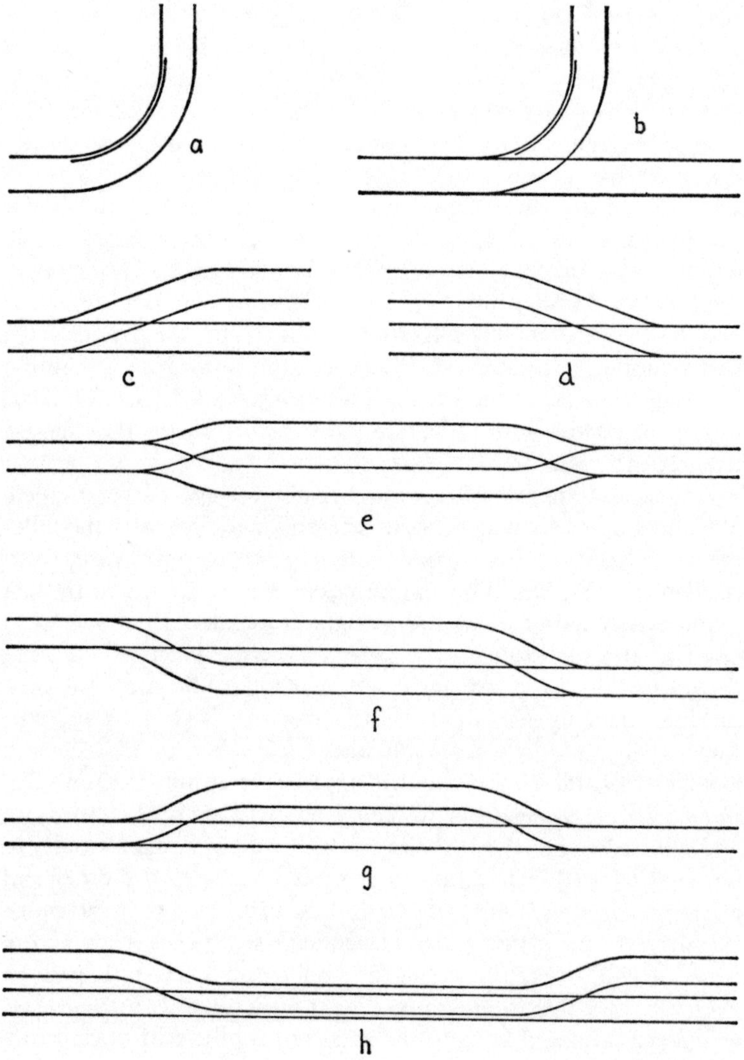

FIG 22 Types of special trackwork

point work must be inserted in the jig for the whole extent of the actual point-crossing. The trailing crossover *c* and the facing cross-over *d* are a little more complicated to construct and it would be advisable to make a special jig for these. The construction of points will be explained but it is essential to describe the types of special trackwork first. A diamond or 'Y' turnout is shown in *e*, and *f* is known as a 'right-hand thrown over turnout' normally seen in restricted space. An ordinary siding or turnout is shown at *g* while *h* is known in tramway parlance as an 'interlacing line' as this type of trackwork was used where double tracks came to a narrow section of roadway. It saved the insertion of two points which are more costly.

By following a few simple basic rules points are comparatively easy to construct and making one's own points is far cheaper than purchasing manufactured ones, especially in the larger scales. The cardinal principle is meticulous accuracy in shaping throughout. First the sleepers, of differing lengths depending on the size of the turnout of the point, are placed in the jig (Fig 23). The straight

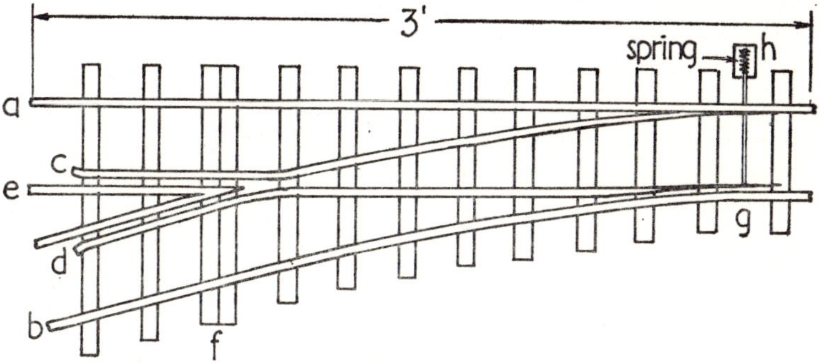

FIG 23 Pointwork construction

rail *a* is laid first but *before* this is done a small section of base-flange near one end is filed down *on one side* only to the level of the web of the rail. This slight recess is to ensure a close-fitting movable tongue. The next rail to be laid is the turnout *outer* rail which must also be slightly recessed to the level of the web on the inside where the tongue of the point will close. When these two basic rails forming the outer layout of the point have been screwed tightly to the sleepers it will be seen that a rigid base has been formed, making subsequent work much easier.

The next section to prepare is *c* which has to be filed down to a gradual knife-edge to form the tongue of the point making a perfect fit, remembering to leave sufficient rail over at the other end to form the check rail at the frog of the point. The corresponding movable blade *d* is likewise filed and bent to shape. *Before* fixing these two movable sections to the sleepers the point-rod (or plate) is *silver*-soldered across the underside of the tongues (*g* and *h*). The point-rod must have a thread cut on the outer end as a spring has to be inserted outside the running rail and held in place by a nut to adjust the tension. All tramway points are spring-points (except in depot roads or controlled junctions) and are self-acting. The car running *through* a point forces it over against the tension of the spring, and as the wheels clear the point it is automatically re-set to enable a car running in the opposite direction *against* the point to take the facing point on the correct rails. The emphasis on silver-solder for point-rods or plates is because it withstands greater stresses than soft solder and with this latter point blades have a tendency to break away and cause derailments.

The two completed blades should now be fixed to the sleepers but it is not necessary to fix them down at the four or five sleepers immediately under the actual tongues as allowance has to be made for sufficient springiness to enable the point to act. The point-rod previously silver-soldered on to the underside of the blades will keep the point true to gauge. The four or five sleepers immediately under the point blades should be equipped with a thin brass strip as, with a little oil on afterwards, the point will slide smoother.

The frog of the point is then filed to form a sharp 'V' section from two small rail sections. In addition to drilling in two places to insert a nut and bolt it is advisable to solder the 'V' and file down the running surface to fine proportions. While laying the movable blades the track-gauge should, of course, be used to verify accuracy. Underneath the actual 'V' of the frog it is also advisable to insert another thin brass strip and solder the frog to it as the more rigid the frog the less likelihood is there of subsequent derailments. It will be seen that extra sleepers are shown under the actual crossing or frog, again to aid rigidity.

The point described is for railway type track used on 'reserved' sections of the tramway. It is possible to imitate tramway street track in the construction of points by inserting the 'L' shaped aluminium check rail throughout, as previously described. In this case the railway check rails each side of the frog would be omitted as the 'L' shaped check rails would provide a neater tramway finish. Additionally the point blades would be made by actually cutting them to form a four-inch movable tongue which would have to be inserted into a correctly machined tongue-box. The four-inch movable tongue is constructed by drilling through the heel of the blade to take a recessed cheese-head bolt and nut to enable the point blade to swivel.

As the construction of points must be carried out with great accuracy, it is better undertaken step by step and each stage of the work carefully checked, when any slight errors can be detected and rectified. Whenever possible it is desirable that crossovers on double track should be laid in a trailing direction so that cars run through them and not against them, another factor obviating unnecessary derailments. It is also important that points should not be constructed with too sharp a turnout. This is always false economy as the snatch or sudden sideways jerk of the car in taking so sharp a point will tend to break the holding-bolts on the bogies of the car bolsters. The more gradual the turnout the smoother the riding of the cars. In 3/4in to the ft scale the ideal is to use a three-foot section of rail for the length of the point from the toe of the point blade to only a four-inch section frog. This will be found to give a car a very smooth glide into the points.

All pointwork must, of course, be bonded with the same care as straight track and in view of the likelihood of imperfect electrical connections at the crossings or frogs and the movable blades of the points, the whole of the pointwork should be cross-bonded by soldering the bonding underneath the actual four rails which open out to give access to the double track.

An aspect which often troubles modellers on outdoor track is the question of the gap to be left between the rails as they meet as rails are subjected to variations in temperature which cause corresponding changes in their lengths. If no gap is left the track will buckle sideways in hot weather and get out of gauge. The modeller will have noticed that on prototype rails a gap of 1/4 to 5/16in is normal in cold weather while on hot days there is scarcely any gap at all. In tramway practice with the rails embedded in the roadway, the surface exposed is small and the rails are not subjected to such rapid or extreme variations of temperature as on open tracks. Quite true, welded track is now more to the fore in prototype practice but without an extensive knowledge of metallurgy it would be unwise for the modeller to attempt it in model tramway track. But both outdoor model tramway track and indoor track, particularly if this latter is laid in lofts or sheds, is subject to temperature variations. Experience has proved that in 3/4in scale a 1/16in gap at all rail joints is sufficient to prevent any expansion troubles.

Trouble is sometimes caused in wet weather on tramway track by water flowing along the grooves of the rails. If this water freezes it damages the track by loosening the joints and it is also a cause of derailments. This can be obviated in tramway grooved track by drilling through the actual grooved rail, at intervals, to allow such water to escape. Prototype tramways placed 'drain-rails' at intervals in places where water gathered. These were about six or eight inches long and consisted of a gap in the groove to allow the water flowing along the rail to drain into a pipe which was connected to the ordinary surface drainage system. Even so, such drain-rails were very infrequent so there is no need for the modeller to go to such lengths, the method described of drilling occasionally is quite sufficient for the purpose of scale model tramway track.

In connection with trackwork a brief account of the conduit system of current collection is given although in modelling its disadvantages are many and it is not to be recommended. It was used in some large cities where objections were raised to overhead trolley-wires, the most notable being Berlin, Brussels, Budapest, Bordeaux, Bournemouth, London, Lyons, New York, Nice, Paris, Vienna, and Washington. The conduit was a circular pipe, open at the top, to which access was gained by a continuous slot in the roadway. The conduit was normally laid between the running rails or (as at Bournemouth) actually under these and the access slots formed in one of the running rails. Tie-bars kept the 'U' shaped conduit in place and each side of the 'U' conduit were installed 'T' bars on insulators to carry the current of opposite polarity. Thus, on conduit systems, track-bonding was not necessary. The current was collected by means of a 'plough' underneath the car which was detachable at various 'plough-change-pits' where the conduit track ended and overhead trolley collection was in force.

The construction of conduit system would present any modeller with immense difficulties, especially at pointwork where insulation and trackwork problems are quite substantial. One moderately successful modification adopted in modelling a scale model conduit system has been to use a modified form of the conduit using the top of the slot rails to act as a conductor and also a modified form of plough accordingly. Even so, the operating difficulties are of sufficient proportions to deter any but the most dedicated.

There was one other method of current collection connected with trackwork which was in use in the early days of tramways and which has been used in a modified form on scale model electric tramways at exhibitions when a prototype steam tramway engine drawing a passenger coach needed to be demonstrated. Obviously a steam tramway engine of the type in use before the advent of electric tramways would look incongruous if equipped with a trolley pole so recourse had to be made to another less visible method. For the first years of their operation some tramway undertakings used what was known as the 'surface contact' system. This consisted of a series of metal studs in the centre of the track, placed at intervals which in appearance were not unlike the 'Cats-eyes' on present-day roads. The tramcar carried a long collector plate or 'skate' slightly above the road surface, which electromagnetically attracted the recessed studs, two or more at a time, as the car progressed. Under the roadway in a conduit was a continuous power-cable in contact with the studs. Immediately the tramcar broke contact with the studs, by the passage of the car, they returned to their 'dead' position by the action of a spring, thus becoming de-energised. Only about six systems used this form of current collection in England and it was finally abandoned in favour of overhead trolley operation because there were occasional instances of the springs failing to return a stud to the neutral position and it remained 'live' with a consequent injury to pedestrians or horses. But for a modeller who contemplates modelling a prototype tramcar using this system (Wolverhampton, Torquay, Mexborough, Lincoln, Hastings, and experimentally in part of London) a description of a modified form of 'surface contact' is given, suitable for scale modelling.

Quite obviously on a 3/4in to the ft scale model operating at 24 volts dc there is no necessity to make the studs electro-magnetically operative as the low voltage used would not justify the excessive time and energy being devoted to the construction of scores of such studs. The 'skate' on the model tramcar is quite easily made from a strip of thin copper, positioned to contact the studs, the resilience of the copper being sufficient to provide its springiness. The studs consist of roundhead brass 3 / 4 x 4 screws or snaphead copper rivets which are installed at correct intervals centrally between the track in the 'street paving'. In practice the distance between the studs is governed by the length of the copper skate underneath the model tramcar as this must always be in contact with at least *two* of the studs at any time to ensure smooth conductivity. The studs are soldered securely underneath the street paving to a continuous length of 18 swg copper cable which is wired up in exactly the same manner as will be described for a trolley-wire feeder. A break in the feeder cable is made at points and crossovers to facilitate control of the cars using this type of current collection. Care must also be exercised in ensuring that the continuous feeder below street level is not allowed to touch the rails at points and crossings.

Construction of depot track and pointwork depends on the space available for the depot and also upon the number of model tramcars it is intended to house. In a remote-controlled scale model tramway, whether indoors or outdoors, it is advisable to install the access track by means of a trailing point. The advantage of this is that, assuming the track is double track past the depot, the depot access point can be a spring point so that it is only necessary to run a car through the points and by reversing it, shunt it into the depot. The added advantage is that facing points without springs, unless they are manually operated with a point-lever or a solenoid, can become displaced and cause a derailment which, at speed, could damage the model.

Having gained access to the depot from the running tracks the number of depot 'roads' and the necessary points giving access to them depends on the requirements and extent of the layout. Both the access track and the individual depot tracks will, of course, have to be individually controlled by separate circuits, preferably by an additional controller. The insertion of ordinary tumbler switches in the circuits between this controller and the relative depot roads will ensure that only the models needed will be made operative, on their respective tracks provided the individual tracks or overhead have been properly electrically isolated.

While some modellers will install a small depot controller near the depot where it is convenient to operate and also turn the manual points, other modellers may wish to remote-control their depots and all car movements using solenoid-operated points. With a system using trolley-poles this can present problems as reversing a car with the trolley-pole thus running in the wrong direction, dewirements can be caused at overhead frogs. If the modeller is modelling a system using pantographs on the cars no reversal problems exist while, with care, models using bow-collectors if driven slowly can also be made to reverse successfully within the confines of a depot area.

10 OVERHEAD WORK AND TROLLEY-WIRE SUSPENSION

The 'overhead system' is a term applied to tramways using a conductor of current suspended within reach of the trolley-pole, bow or pantograph carried by the tramcar. If the conductor is immediately over the centre of the track it is described as 'centre-running' but if the conductor is some distance from the track centre, it is known as 'side-running'. The centre-running system is the most usual and mechanically correct and the side-running system was only adopted when local authorities raised objections to the unsightliness of the centre-running one in the early days of this century, although what attitude would be adopted in recent years towards tramway wires, having regard to the cobweb of electricity cables in streets, is a moot point. The side-running system is useful for short distances underneath bridges where sufficient clearance is necessary for the tramcars but, even in these cases, the distance of the trolley-wire from the centre of the tracks was never more than eight feet otherwise the trolley-pole would leave the wires.

Various types of street poles are used. Steel poles are the most common in England although wooden poles were used on the Kinver Light Railway and they were common in America, and on the Continent of Europe. Lattice poles are used in some cases on the Continent but the normal tubular poles used in general are made in three sections tapering towards the top. In modelling work this is achieved by using tubing of different sections and inserting one inside the other, alternatively the poles and (in some cases) the ornate collars can be machined on a lathe from brass rod of an outside diameter of 1/2in. The amount of detail and decorative work involved depends on the time the tramway modeller is prepared to devote to this sphere of his layout.

A simple tubular pole can be made from ordinary gas-piping of an outside diameter of 1/2in throughout. This is quite cheap and, for 3/4in to the ft scale can be cut into the necessary 1ft 8in sections

The simplest method of suspending the trolley-wires is to hang them from a span wire suspended between two poles on each side

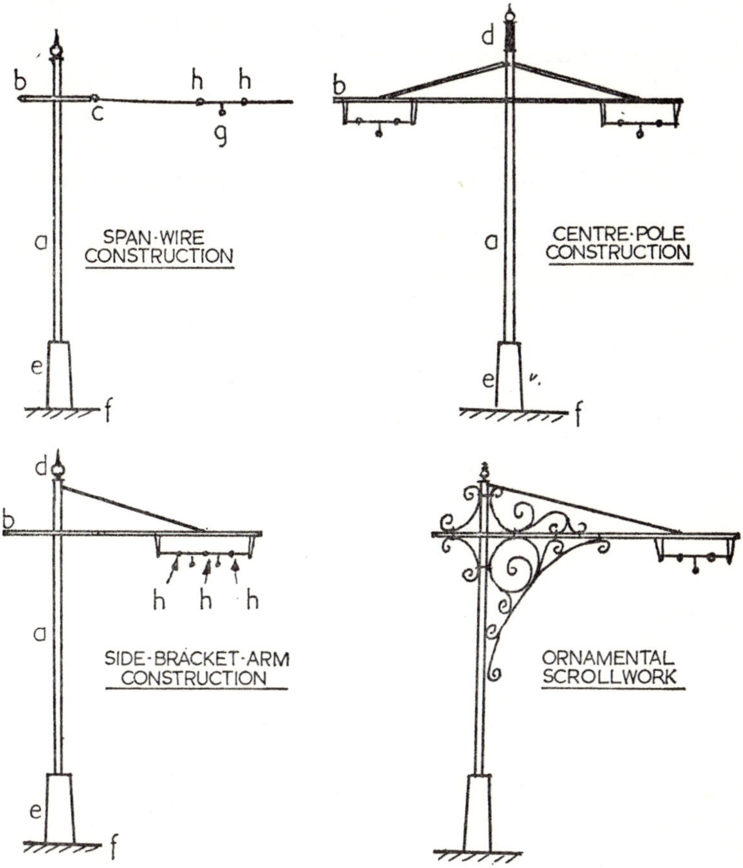

Fig 24 Trolley-wire suspension

of the track, known as 'span-wire construction'. The illustration Fig 24 shows the arrangement for centre-running on double tracks, in half section, the other half being similar. The pole *a* is forced into the larger section tubing *e* which is soldered on to a base-plate *f*. This latter is cut 2in by 1in by 1/4in from brass plate or bar and drilled to take two screws for fixing to the baseboard of the model. The bracket *b* is made from small-section Calor-gas tubing and the pole is drilled to take this. A small swan-neck hook upon which to affix the span-wire is soldered in at *c*. Two insulators are inserted in the span-wire at *h* and these can be made from two-hole pearl buttons which can be filed down to represent the correct pattern tramway insulators. The trolley-wires are then hung from a hanger at *g*. In modelling the hangers are cut from very thin brass sheet to encircle the trolley-wire and then soldered on to the span-wire at the top.

In cases where the model layout is laid directly on to the garden without any baseboard, the poles have to be at least twice the height mentioned and then driven into the ground accordingly, the base-plate being dispensed with.

Another method of suspending the trolley wires is the 'centre-pole' construction. Although centre-poles were used in many cases in the early days of tramways, their use was abandoned latterly, as ordinary road traffic became more intense, and span-wire construction substituted. Centre-pole construction is, however, widely used on 'reserved' or private right-of-way sections of track. They are made in exactly the same manner as the span-wire poles except that the bracket *b* is of longer section Calor-gas tubing which is easily heated and bent round at the ends. These ends should be flattened with a hammer and drilled to take the span wires. The inside brackets (nearest to the pole) are similarly made from flattened Calor-gas tubing which is then soldered on to the cross-arm, after being drilled for the span-wires. Insulators are, of course, inserted as usual in the short span-wires each side of the hangers. The supporting beams from the top of the poles to the cross-arms are made from welding-rod which can be obtained from any garage. The pole is drilled through the centre to take this rod which is then soldered on to the cross arm at each end, afterwards being filed and buffed to give a neat finish. Side-bracket construction follows the same principle, remembering to insert an extra insulator between the two trolley-wires, at *h*.

The use of metal scroll-work depends on the inclination of the modeller. It was extensively used in the early days of tramways as, in an endeavour to meet local objections regarding alleged

Page 107 (above) Blackpool Corporation single-deck railcoach No 319 and double-deck car No 257 passing each other at night on the Glastonbury layout; *(below)* two 3/4in to the ft scale Metropolitan tramcars running on an exhibition track in London. An open-topped early type car No 171 passing a modern 'Feltham' type car No 359

Page 108 (above) 3/4in to the ft scale model of London Transport class E/1 car No 1574 by Mr R. Elliott, shown at the National Models Exhibition in London; *(centre)* depot car No 020 (ex-LCC) permanently fitted with snow-plough and service car No 2055 of London Transport (ex-Walthamstow) assisting, on 3/4in to the ft scale garden tramway; *(below)* exhibition stand of the Tramway & Light Railway Society at Central Hall, Westminster, with 3/4in to the ft scale models belonging to members of the Society. Both narrow-gauge ($2\frac{5}{8}$in, foreground) and standard-gauge ($3\frac{1}{2}$in) models are being demonstrated

unsightliness, it improved the street furniture. In later years, because of rusting of the collars which held the scroll-work in place, it was removed, on most tramway systems. It is, however, easily made from 1/8in brass strip which is soldered on to the cross arms and supporting beams.

At the top of each pole of whatever type is a *finial, d,* and these varied in shape according to the distinctive pattern adopted by the relative tramway system. In modelling they can either be turned in brass on a lathe or cast in lead in a plaster-of-Paris mould, in each case remembering to leave sufficient base to the finial to insert into the top of the pole after first applying Bostik to it. All poles, bracket-arms, supporting beams and (if fitted) scrollwork should be painted in the appropriate colours of the system being modelled. Green was a fairly standard colour although some systems favoured aluminium paint while, for a time, the Black Country systems were painted brown. If the layout is out of doors, the poles should be painted annually.

An important point to remember when constructing trolley-wire suspension is that if the modeller is building his models with either bow-collectors or pantographs the hangers suspending the trolley wires must be deeper than for trolley-boom collection. The trolley-wires must hang quite clear of the nearby supporting brackets, slightly lower than normal, otherwise the contact plate of the bow-collector or pantograph will foul the supporting brackets. Also in hanging trolley-wires for bow or pantograph collection, it is customary to zigzag or stagger the trolley-wire slightly so that the wear on the conducting surface of the collecting plate will be distributed, otherwise the wire will cut a 'V' in the collecting plate.

The distance between the poles in the direction of the track was 120ft in actual practice which, for 3/4in to the ft scale modelling would be 7ft 6in apart. On curves and where there is special overhead wiring (as entering depots or junctions) the poles would have to be set closer.

On curves, where it is necessary to ensure that the trolley-wire is maintained centrally over the track the use of 'pull-offs' is necessary and these do not take the weight of the trolley-wire but only the side strain. The positioning of the poles is shown in Fig 25 at *a, b, c,* and *d.* The span wires between *c* and *d* and *d* and *b* are normal but the pull-offs at *f,* four of them attached to pole

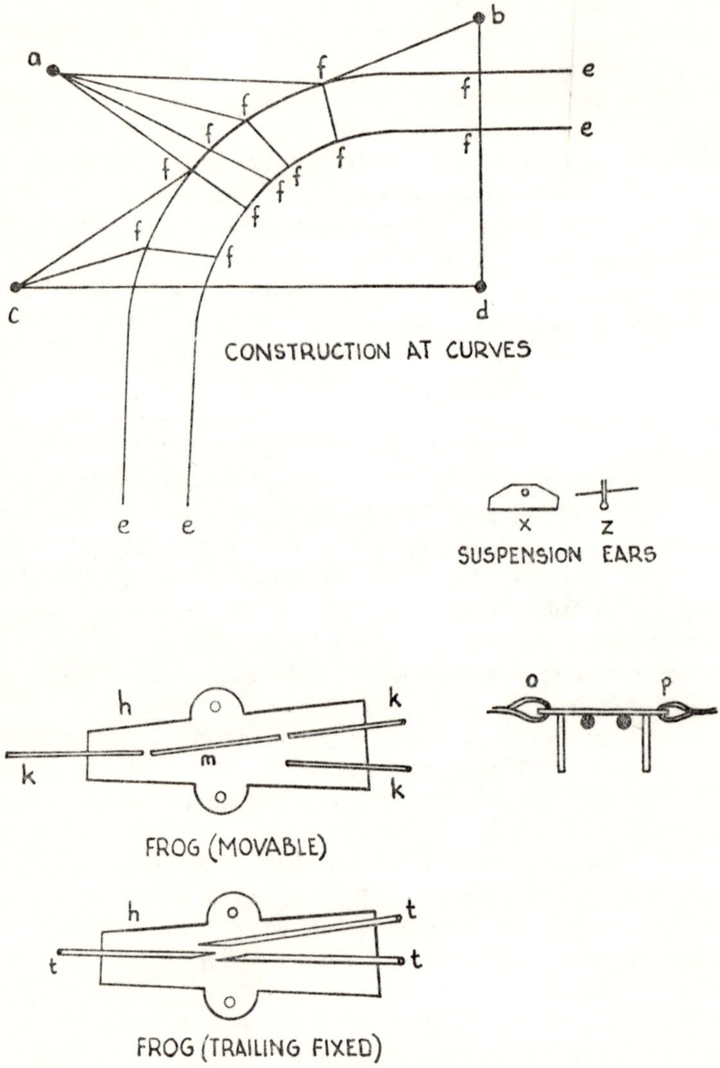
a
b
c
d
e
e
e
e
f
CONSTRUCTION AT CURVES
x
z
SUSPENSION EARS
o
p
h
k
k
k
m
FROG (MOVABLE)
h
t
t
t
FROG (TRAILING FIXED)

Fig 25

a, two to pole *c*, and one to pole *b*, indicate how the side strain is taken. The correct positioning of these pull-offs will obviate any dewirements as the trolley-wheel will negotiate the curve smoothly. If insufficient pull-offs are used, the trolley-boom will jerk, or snatch and this itself is a constant cause of dewirements. It may be found that, in order to maintain the correct tension of the trolley-wire on a curve, the anchoring of the pull-offs at pole *a* might have to be slightly higher than normal but only trial and adjustment will ascertain this. The pull-offs must, of course, be insulated throughout from the supporting poles and from each individual trolley-wire exactly as described earlier in connection with the general principles of overhead installation.

In order to support the trolley-wire in position, small fittings called 'ears' are attached to it at the span wires. The usual form of ear is shown in the sketch at *x* and *z*. These are cut from 1/32in brass sheet with a pair of metal snips and shaped accordingly. A hole is drilled through the top, to take the span wire and, after the ear is clipped round the trolley wire and pinched tightly to shape with combination pliers it is *held in that position* with the pliers while the span wire is soldered to make the joint at the hole. The type of trolley wire used for outdoor model tramways (which can also be used for the span wires) is 18 swg although a finer gauge wire can be used for indoor layouts. It will be necessary to solder very lightly the ears on curves as, with the passing of the trolley-wheels of the cars, they have a tendency to move their positions. The underside of the ears round the trolley-wire should be trimmed lightly with emery cloth in order that the trolley-wheels may have an uninterrupted course when passing underneath.

18 SWG = .048
4mm SCALE : .010
or 33 SWG

At junctions, overhead switches or 'frogs' are used to guide the path of the trolley-wheels from one wire to the other. In modelling these frogs are made from 1/16in sheet brass using trolley-wire soldered to the underside of the frog. The illustration shows a movable tongue at *m* to guide the wheel to the proper wire. The tongue is swivelled and operated by a suitably insulated steel wire carried down the nearest pole where it can easily be pulled by the operator to the junction required. The trolley-wires entering the frog are shown at *k* and it is held in position by the span wires at *o* and *p*, attached to the frog at *h*. Much ingenuity in operating movable frogs has been evidenced by tramway modellers on their layouts. In some cases the frog is worked automatically by the trolley-pole of the car going round the curve which operates a lever to move the tongue, as was done on some prototype systems in actual practice. In other cases modellers have installed a remote-controlled solenoid to operate the tongue. Some modellers have fitted ingenious springs to the movable tongues to enable them to be held over for the subsidary line while returning, by the spring, to the main line as normal.

The illustration at the foot shows the construction of a *fixed* frog, also known as a trailing frog. This type is used where the trolley-wheel has to pass from a branch to a main line in a *trailing direction only*. There is, in this case, only one path the trolley-wheel can take. By whichever wire the trolley-wheel enters the frog on the right, it can only pass to the main line and all trolley-wires are shown as *t*.

When installing frogs a great deal will depend on how they are inserted. Their efficiency can only be ascertained by trial and adjustment so it is advisable not to have the trolley-wires too taut in the first place, leaving some latitude for adjustment. It is often necessary to adjust the frog after installation, usually by tightening the span wires.

Section insulators inserted in the trolley-wire are used to divide and insulate it in required sections. In prototype practice in England they were required by law to insulate sections at half-mile intervals. In modelling they are essential to sectionalise various sections of the track whereby, by remote control, cars may be held and, therefore, better and more realistic operation is possible. They should *always* be inserted on *one* of the trolley wires entering a frog (usually at the subsidiary side of the junction) to insulate the minor road from the main line.

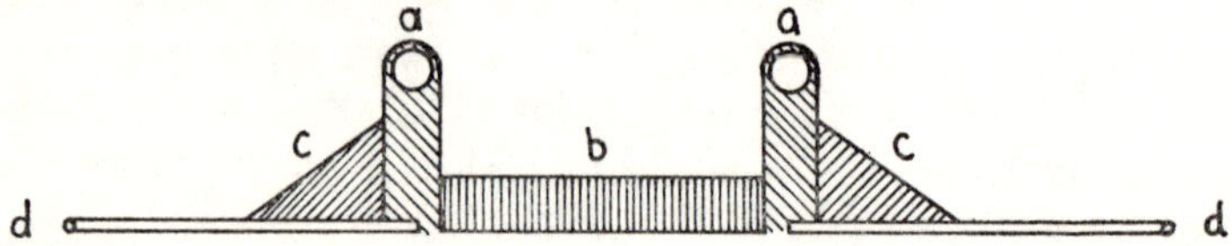

FIG 26 Section insulator

A simple type of section insulator is shown in Fig 26. The trolley-wires *d* are attached to the two 1/8 by 1/4in brass strips into which have been drilled holes at the top *a* for suspending on the span wires. The connecting beam *b* is a compressed strip of hardwood planed down to form a 'V' for the passage of the trolley-wheel underneath. The reinforcing strips *c* are brass sheet soldered firmly on to the trolley-wires to give rigidity. The hardwood insulating strip is affixed to the two end brass sections either by cutting small slots in them and inserting two 10 BA bolts and nuts at *each* end to prevent lateral movement or by making two small 'L' brackets from brass strip, soldering them on to the upright brass strips *a* and drilling them to take two minute brass screws at *each* end of the hardwood. There are many types of section insulators and, here again, is scope for the modeller's ingenuity.

As the trolley-arm has to be turned round and changed from one wire to the other on reaching a terminus so that it trails behind the car a useful adjunct to any model tramway is a 'trolley-

FIG 27 Trolley-arm reverser

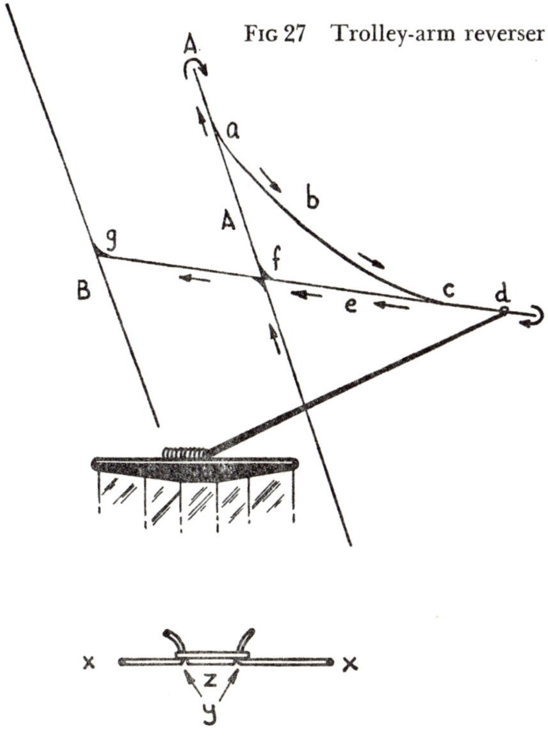

arm reverser', Fig 27. This is found especially useful when the models are operated by only one operator without an assistant as, with remote control, it is not then necessary to keep going from one end of the layout to the other to reverse trolley-arms. The trolley-arm reversers used in prototype practice worked on the principle that a car arriving at a terminus (towards the reader) uses line *A* throughout. When the car is reversed it pushes the trolley forward until it passes over *a* in the opposite direction, the trolley-wheel then taking the wire *b* over the frog *c* to wire *d*. The trolley-arm is then at its fullest extent at right angles to the car. The car still moving forwards, the trolley-wheel still on *d* approaches frog *c* again which, this time, diverts it to wire *e*. The wheel then proceeds under the frog *f* and through the trailing frog *g*, enabling it to join wire *B* correctly aligned in a trailing position. The arrows show the path of the trolley throughout. Only the actual necessary trolley-wires have been shown in the illustration for the sake of clarity but, of course, the relative span wires for suspension of the trolley-arm reverser have also to be worked out. Again a section insulator is necessary between *g* and *f* to isolate the 'up' and 'down' circuits. While trolley-arm reversers may sound complicated, they really are quite simple to make and, if constructed with care, give no trouble—provided that cars are driven through them at slow speeds. At several Model Tramway Exhibitions where they have been installed they have been effective throughout the whole week of the exhibition without a single dewirement.

In erecting trolley-wires tension is most important. Although prototype tramways allowed for a 45°C range between summer and winter temperatures which corresponded to a variation of nearly 4ft per mile in a trolley-wire's length, obviously this would not apply in a model tramway. At the same time some slight allowance must be made for sag particularly on outdoor model tramways, otherwise contraction in winter will subject the copper trolley-wires to acute strains and can, in fact, distort the whole of the overhead construction.

A simple method of joining trolley wires (where frogs have been inserted for example) is the one illustrated, which requires no soldering. A small brass strip 1/2in by 1/8in by 1/8in is drilled at *y* with a fine drill. A small section of trolley wire is soldered on the under side at *z*. The two ends of the trolley wires *x* are then inserted through the drilled holes and bent back at the top to make secure.

11 OUTDOOR MODEL TRAMWAY PLANNING

This aspect of a model tramway depends upon the conditions of the site. If the garden is perfectly level the line can be planned round the boundary fences, raised to a height of about 3ft on posts or trestles. It can also be fixed to boundary walls by substantial shelf brackets, but if the garden is surrounded by hedges or light fencing it will have to be constructed on posts. It is always advisable to have the line at a height of about 3ft as continual stooping to lay track, construct overhead trolley wires, and attend to model tramcars will be found to be tedious. Should the garden be on a gradient it will be necessary to construct the line at ground level, or even in a cutting at one end, rising on posts as the gradient falls away. This is necessary because in a model tramway the line has to be perfectly level for operative reasons. For instance, difficulty will be experienced with the models if they are expected to tackle gradients as the small motors used in the normal 3/4in to the ft scale are not sufficiently powerful for this purpose. There is also the consideration that on gradients, as the models are remote controlled, a dewirement can cause the operator to lose control and the model will run away with consequent damage.

The main essential is a firm foundation for the track and on this depends the entire success of the layout. If supporting posts are to be used they should be of substantial proportions, at least 15in in circumference, well creosoted and embedded in a 2ft foundation with a concrete base. They should be at intervals of not less than 3ft with a timber 2in by 2in cross-member to form a 'T' securely nailed on top (Fig 28). The actual shelving on which the track is to be laid should consist of 1¼in planking which, for double track, should be constructed to a width of at least 18in in

3/4in scale. The shelving should be given a coat of creosote and, when dry, roofing felt must be affixed to the top, overlapping at the sides and nailed with the excess overlapping pieces *on the undersides* of the shelving with large-headed aluminium felt-nails.

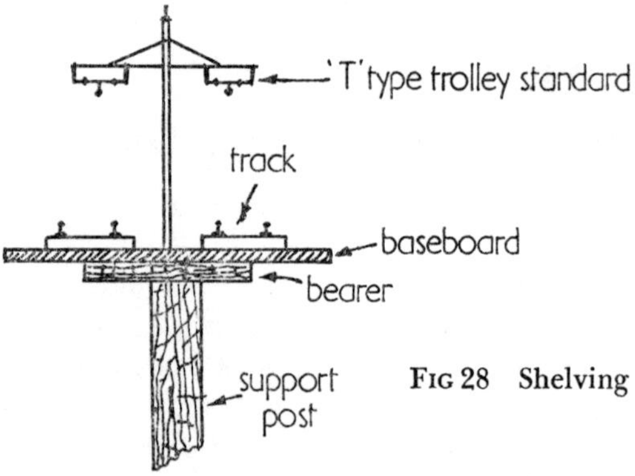

FIG 28 Shelving on supports

Unless the framework and boards are laid to this specification the subsequent cracks and warping will be the cause of endless trouble.

The necessity for constructing the shelving to a width of 18in for double track will become apparent when the track is laid and the overhead trolley-wires have to be installed. It will be found that in $3\frac{1}{2}$in gauge the outside edges of the sleepers extend to 14in so it will be seen that the remaining 4in is necessary for the positioning of the supporting trolley poles. If span wire construction is favoured the clearance in this gauge will be sufficient but, for a garden model tramway bracket arm suspension is preferable and on the 'reserved' private right-of-way section of the tramway centre 'T' poles are much to be preferred. Both bracket arm and centre poles allow easier access to the tramway. It is, however, a matter of preference for each individual modeller.

When installing the track, previously constructed in the workshop, it is only necessary to drill about every tenth sleeper through the centre and screw to the baseboard with $1\frac{1}{4}$in (size 8) *brass* screws. Under no circumstances must steel screws be used as they rust and any subsequent realignment of the track or adjustment will present difficulties. When positioning pointwork practice will indicate what additional fixing with brass screws is necessary.

If, due to a gradient, part of the track has to be laid directly on to the ground, it will be necessary first to excavate the soil to a depth of 9in by 14in and tread down the base of the excavation firmly, filling in to a depth of 8in with ashes or clinker and also treading this down firmly. The remaining 1in is filled with ballast, and the *small* size chippings used as top-dressing on macadam road surfaces during the summer will be found most suitable for scale ballast. The track is now installed with the aid of a spirit-level the ballast being packed round the sleepers accordingly. When all the track is firmly in place the track joints are bonded with 18 swg copper wire securely soldered at each end.

Now it may be thought that such close attention to a firm 'road-bed' has been over-emphasised but experience has proved that there is nothing more discouraging than to discover that track has become distorted or out of alignment. In the case of track laid at ground level, attention to drainage is of vital importance to obviate flooding.

The erection of the overhead trolley-wires now claims attention and foresight in allowing a wide baseboard will be rewarded. In 3/4in scale the poles should be positioned at 7ft 6in intervals but it may be found that, in modelling, they need to be closer to prevent wire sag. They should be firmly screwed to the baseboard, again with $1\frac{1}{4}$in brass screws and the copper trolley-wire installed over one track first. When the suspension ears have been clamped into place it is advisable to test run a model tramcar to observe whether the correct tension has been achieved, adjusting either by tightening or slackening accordingly. The second trolley-wire can now be installed with the previous one acting as a guide. At night-time it will be noticed that the trolley-wheel sparks under the ears; this is because they are new and have not yet 'settled' or worn in. The sparking will disappear as they settle.

After a few weeks when the track has endured some rainstorms followed by dry spells it will be necessary to re-align the track at ground level with a spirit-level and re-pack the ballast in places. Provided this is carefully done, it will only need such attention about twice annually in future. If a model tramcar sways, at any part of the track, an investigation should be made as it will be found that the track needs realigning there. On an outdoor layout more derailments are caused through incorrect track alignment than gauge distortion as a check with the spirit-level will often prove.

A question frequently asked in connection with an outdoor layout is about its durability. If the model tramway is originally constructed to the standards set out in this book there is no reason why it should not be operative for very many years and, here again, it is surprising what little maintenance is required to keep it working satisfactorily. Some members of the Tramway & Light Railway Society have operated outdoor model tramways for twenty years. Provided initial attention is paid to insulation requirements, neither damp nor storms will affect the permanent way or the overhead trolley-wires. These latter are impervious to high winds and storms as, being of such small proportions, they offer little resistance. Ice on the trolley-wires does not have the same effect as similar weight on telephone wires as on a model tramway the trolley-poles are much closer and give adequate support in such circumstances. If it is intended to operate the model tramway in these inclement conditions all that is necessary is to spray the overhead trolley-wires with a solution of common salt, treating the points similarly. The construction of a snow-plough is well worth the trouble spent in making it as it is quite simple to construct. Provided the models are soundly constructed the tramway can be operated throughout the years in all weathers. The 24-volt motors described for 3/4in to the ft scale are extremely robust and, being sealed, are not affected by weather conditions. If the track is laid at ground level the backbreaking task of weeding the ballast is easily obviated by a solution of sodium chlorate mixed in a watering can and sprayed on the track in early spring and perhaps again in late summer. The resultant sparking from the wheels of the cars for the next few days is quite entertaining, especially at night.

A problem confronting model tramway engineers in a garden layout concerns the crossing of a garden path. If the line is constructed to a circular route (one most favoured as it entails no trolley turning) it will have to cross the path leading from the house to the garden. This presents no insuperable difficulty as two methods are open to the model engineer. The first and most obvious is by building a small concrete overbridge with steps or a ramp on either side. The minimum clearance required for a 3/4in to the ft scale model double-deck tramcar is 15in to include the trolley-boom and trolley-base on the top of a top-covered car. The overhead trolley-wires are conveyed under the bridge in ducts or troughs which are easily made, in wood, with the suspension hangers affixed directly to the underside of the ducts. It would be a poor gardener who could not raise his garden path a mere 1·5in for so worthy a purpose. The second method of crossing a garden path is by means of a Scherzer Bascule bridge, a very successful type of lifting bridge. A modification of this type of bridge for this purpose would entail using special hinges to ensure that the tracks were lifted clear of their permanent counterparts on the hinged section. The overhead trolley-wires also need reception troughs at the opposite end to the hinged section as, with the breaking of the trolley wire, this method of preventing dewirements when the bridge is in use is advisable. A number of Scherzer Bascule bridges are used on tramways in Holland but with the bow and pantograph collection of current on their tramways, the question of dewirements is practically non-existent as deflector wires are provided at angles overlapping the normal trolley-wires at each end of the bridges. In the case of reception troughs to ensure that the trolley-wheels did not dewire at the opposite end to the hinged section of the bridge, a study of Brussels tramcars emerging from their subways provides an examples of these reception troughs. In the subway section the cars use pantographs and, at street level, trolley-booms; these latter re-wire themselves with the assistance of reception troughs as the cars emerge from the subways. If a lift bridge is the only practicable method of carrying the model tramway across a garden path it has to be borne in mind that electrical continuity must be maintained by bonding the 'broken rails' with cables laid underneath the pathway to join the permanent track at each end. The hinged section of track will also require bonding with flexible cable. The same principle applies to the trolley-wires: a continuous cable soldered on to each end of the permanent sections on either side will, if taken underneath the pathway, ensure electrical continuity. It is not advisable to rely on temporary clips to hold the broken trolley-wires together.

The illustration, Fig 28, shows a typical construction in a garden with the supporting post and cross beam underneath the track base upon which is laid the 'reserved' section track. The centre type trolley-pole of the 'T' type is commonly used on this form of track but side poles with bracket arms can also be used if desired as can double poles with span wires.

A popular type of layout is illustrated in Fig 30, showing a circu-

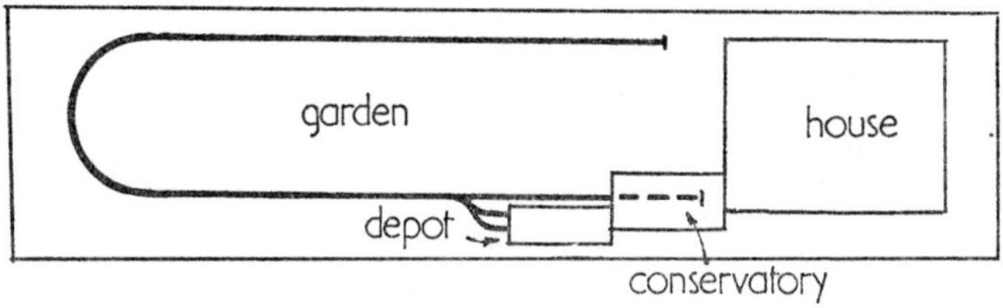

FIG 29 Outdoor terminus layout

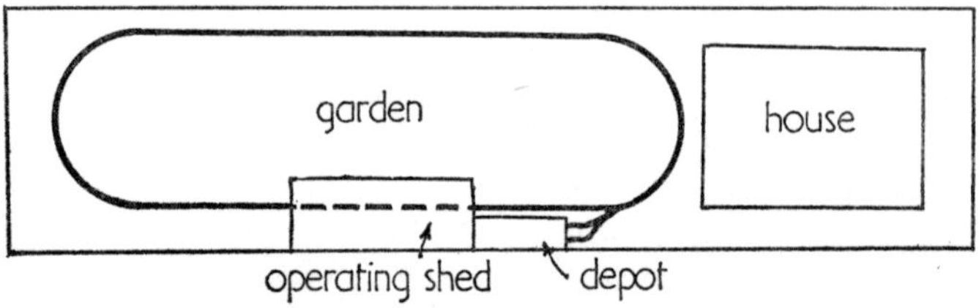

FIG 30 Outdoor circular layout

lar track running through the 'tramway operating shed' in which are installed the controllers, and the nearby depot for housing the models. Another type with a combined indoor and outdoor layout is also shown (Fig 29), which can be operated from indoors in a conservatory or shed, with the depot also nearby. In this case no crossing of a pathway is necessary but the far terminus would need a trolley-arm reverser unless bow or pantograph collection was used. Large radius curves are always advisable and this is where a garden model tramway always scores over an indoor line with its restrictions.

In planning an outdoor model tramway due consideration must be given to the siting of the depot, tramway waiting rooms, cross-overs (on the double track sections), passing loops on the single lines and, on the 'street' sections of the tramway, street lighting. The size of the depot will depend on the number of cars it is intended to house. Scale model depots are easily constructed with timber sides and a galvanised sheet steel roof. A realistic tramway-type cowl-shaped apex for the whole length of this roof can be obtained from any agricultural-house stockists as it is of the same pattern as those used in fowl-houses. The siting of the depot will depend on the space available at a convenient point in the garden having due regard to its accessibility. The number of tracks leading into the 'depot roads' is also a matter of choice as they all converge on one single track to join the service lines. The ancillary buildings surrounding a tramway depot are again a matter of choice and some tramway modellers devote considerable effort to constructing inspectors' offices, a depot superintendent's office, traffic office, staff canteen, ticket office, and similar buildings which, of course, all add to the realism of the layout. The lighting of all these ancillary buildings, the street lights, and the waiting room lights (on a separate circuit to the traction current) is most impressive at night. Some tramway modellers have installed pedestrian crossings with flashing amber lights, tramway signal lights in red and green where passing loops are sited awkwardly, police boxes with a flashing blue light surmounting the top, scale post-office pillar boxes on the 'street' sections of their models in addition to a replica of a roadway under repair protected by miniature red danger lamps. The amount of impressive detail possible is almost endless on an outdoor model tramway.

If it is intended, for appearance sake, to mask the posts or supports for the line or parts of the substructure, the ground below can be planted with privet bushes and, when these are fully grown, the tops are trimmed to the surface level of the tramway, leaving here and there a few stems above the rest to give the effect of model trees adding to the scenic effect.

12 CONTROL AND EQUIPMENT

The power supply to a model tramway requires some thought beforehand, for nothing is more frustrating to the modeller who has constructed a worthwhile model or models than to discover that he has purchased inadequate equipment, resulting in unsatisfactory running. The golden rule is to look to the future and calculate for additions to the rolling stock by the construction of more models because it is rare that, having constructed one model tramcar, the modeller is satisfied until he has added to his fleet. By purchasing an adequate power unit in the first place, although the cost is greater, subsequent disappointment is eliminated.

While the advertised 'power-packs' for the smaller-gauge model railways may be suitable for the equally smaller model tramway systems and the controllers also supplied for operating them will likewise give satisfaction, a more substantial supply is needed for any 3/4in to the ft or similar scale model tramway, whether indoor or outdoor. Electric tramcars are driven exclusively on the *direct-current system,* the direct-current motor being ideal for traction work as its characteristics include a variable speed, ample starting torque, and decreasing torque with increasing speed. The 24-volt motors described throughout this book are, therefore, perfect for model tramway operation.

The most satisfactory and trouble-free method of supplying power to a model tramway is from the public electricity mains and this is the one most commonly used by tramway modellers. As the public supply is mainly alternating current, usually at 240 volts ac, it has to be transformed down to 30 volts ac and then rectified to 24 volts dc, to operate the model tramway. In these days plenty of suitable transformers or transformer-rectifiers are available from electrical equipment manufacturers or from the government-surplus shops which specialise in electrical equipment. The overriding consideration in the purchase of the unit is to ensure that the output is adequate. The output for an outdoor model tramway should be not less than 20 amperes although it may be lower for an indoor line. A little thought will show the modeller why a high output is necessary. The scale model tramcar usually takes 2 amperes on starting, falling to 1½ amperes when it is 'under way'. A full complement of lights on the model accounts for another 2 amperes thus, at night, when the lights are on the model will require 4 amperes to start, recovering to 3½ amperes when running. The street lights, waiting room lights, depot lights, and other auxiliary lighting on the layout also have to be accounted for so, with a fleet of only four cars, it will be seen that the 20 amperes stated is only just adequate. A lower amperage will only result in sluggish starting and running and dull lighting, especially on an outdoor layout. It can be very disappointing when, having constructed a sizeable running track, it is discovered that the power output is insufficient and an additional power unit has to be purchased; moreover, two single power units can be more expensive than the larger unit purchased in the first place. Each tramcar takes only just as much current as is required to drive it.

In very remote districts where no public electricity supply is available, two 12 volt car batteries coupled *in series* to give an output of the required 24 volts can be used. If this method of supply is used where there is a public supply, the batteries can be charged from a normal battery-charger through the mains. Many of the early tramway systems were entirely self-contained and had their own power stations. Some tramway modellers have emulated this and have constructed a complete boiler-house, coal or oil fired, the steam engine driving a 24 volt generator which, in addition to supplying power to the tramway, also charges up batteries for use when the power station is not running. As a matter of interest, some prototype tramway systems likewise used battery-houses for stored power which was used very early in the mornings and very late at nights when the power station had shut down. It was also useful in the event of a breakdown. This type of current supply is mentioned because amongst modellers the emphasis on various aspects of a model tramway system differs, some modellers have a greater interest in vehicle construction, others in track layout, while still others like faithfully to copy the original tramway systems throughout, even to the construction of the power station. But in later years the tramways drew their current from the electricity grid supply so the method for the majority of modellers is the one described at the outset.

The wiring of a model tramway system is quite straightforward and the illustration (Fig 31) shows an elementary simple starting

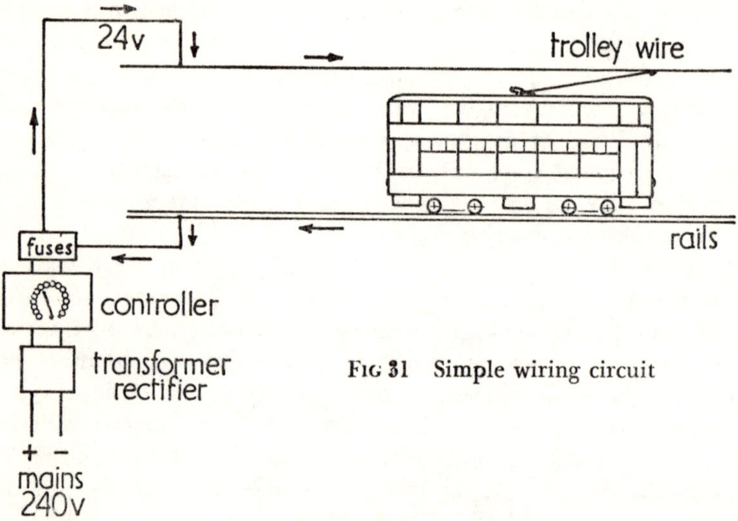

FIG 31 Simple wiring circuit

point. An ordinary mains plug from the house supply is wired to the *input* side of the transformer-rectifier and the *output* wires from the transformer-rectifier are then taken on to the controller which, in turn, is protected by fuses on the output side. One wire is then taken to the trolley-wire of the layout and the other wire to the rails, both being securely soldered thereon. Although I have mentioned fuses on the output side, to protect the equipment from any fault which may develop, a more desirable method is a cut-out switch which is pre-set to operate at a given level. These cut-out switches are also obtainable from government surplus stores and they are entirely automatic in operation. The advantage of a cut-out switch is that it is easily re-set by the switch-arm and, if the fault persists, it will immediately interrupt the circuit again each time until the fault is located, whereas fuses are more trouble to replace. But whichever method is preferred there *must* be either a fuse or a cut-out across the current-supply to the layout.

Page 125 (above) Birmingham Corporation bogie tramcar No 700 emerging from the depot on Mr Robert Whetstone's $1\frac{1}{2}$in to the ft garden scale model layout at Bromsgrove; *(below)* a freelance model railcoach, based on modern tramway rolling stock now in service in Europe, America, and elsewhere. This 3/4in to the ft scale streamlined car was constructed by Mr Peter Hammond while serving with HM forces in Cyprus. It incorporates all up-to-date features of present-day tramway practice including 'PCC-type' bogies, rubber suspension, frame-mounted motors, and cardan-shaft drive

Page 126 (above) Bristol Exhibition with five of the fleet of ten 3/4in scale model tramcars from the Glastonbury model tramway being operated. The souvenir prototype destination blinds were supplied by various former tramway undertakings; *(below)* Manx Electric Railway open 'toast-rack' bogie tramcar No 35 constructed in 3/4in to the ft scale by Mr Peter Hammond

Useful ancillary equipment in the power supply is a voltmeter on the *mains* side of the transformer-rectifier (these are usually made to give a reading from 0 to 300 volts), another voltmeter on the *output* (ie low-voltage) side of the transformer-rectifier (obtainable in readings from 0 to 30 volts) and an ammeter (reading 0 to 30 amperes); this is also wired on the *output* side of the transformer. These three useful adjuncts are also reasonably priced at ex-government stores. The two voltmeters are wired in *parallel* with the supply and the ammeter is wired in *series* as the two former give a constant reading immediately the current is switched on while the ammeter will give a reading only when the controller is operated and a model is running. The usefulness of these voltmeters and the ammeter is that, if the cars seem to be running sluggishly, a quick reference to the voltmeter on the mains side may show that the public supply is 'down' and not being maintained at the proper voltage. If, however, the public supply mains are registering their correct voltage and the output voltmeter shows a lower than normal reading then the fault lies between the output side of the transformer-rectifier and the low-voltage voltmeter. The ammeter, on the other hand, will give an abnormally high reading if, when the controller is being operated, there is a fault in a car or on the track, which may not be enough to trip the cut-out or blow the fuses yet may be sufficient to cause damage. Most tramway modellers regard the foregoing two voltmeters and the ammeter as indispensable, as they save a considerable amount of time in finding occasional faults.

Throughout this book the aim has been to assist the normal everyday modeller who wants to construct a workable scale model tramway within his scope, with readily obtainable materials, at a low cost and with a minimum of technical knowledge commensurate to his capabilities. Even a few years ago, some of the equipment I have described would have had to be made by the modeller but such is the advance of technology that it is now easily procurable commercially. Books on modelling written at the beginning of this century describe in detail how to make one's own rectifiers and cut-out switches described earlier in this chapter. In these days there is very little saving, if any, in going to the trouble of constructing such equipment one's self. This also applies to the next item of tramway operation equipment, the controllers. Prototype tramcars were driven from the driving platform by controllers but in 3/4in to the ft and similar scale modelling this is not possible and the models are remote controlled. A controller is still necessary but, in modelling, it is placed at a convenient point on the layout from which position the operator can see the models he is controlling. A variety of graduated controllers, suitable for model tramways, can be purchased, again from ex-government surplus stores. Many tramway modellers use these controllers on their layouts. They give very satisfactory service and some, on model layouts, have been in use for very many years.

All the same, many modellers impressed by the controllers used in actual prototype tramway practice, are not satisfied with the general small type of controllers available from ex-government surplus stores and insist on constructing their own controllers as near to the prototypes as possible. For their guidance a description of an actual tramcar controller is given.

The whole object of a controller is to feed the current to the car motors from the 'off' position through graduated 'notches' to the maximum power available, by increasing or decreasing the resistances across the circuit. From the 'off' position, moving the controller handle in a clockwise direction, the resistance is gradually cut out until full power is reached when there is no resistance in the circuit.

In the illustration, Fig 32, the main principles of a tramcar controller are shown. For simplicity's sake in order not to confuse the modeller the blow-out magnet, movable insulating partition, and other sections (which are not necessary in a model controller operating on a low 24 volt circuit) have been omitted. The general features are *A*, the controller cover shown open, *B*, the controller handle, *C*, the key which operates the 'forward' and 'reverse' positions of the reversing drum *G* and also immobilises the controller when removed, *D*, the main drum to which are attached the contact segments which make contact with the fingers *H* when the controller handle is moved in a clockwise direction to apply power. The main drum or power cylinder *D* has a *star wheel* fitted at *E* attached to the shaft to ensure that definite positions are 'notched-up' on the controller. This star wheel engages with a spring-actuated roller F to register the definite 'notches' (hence the well-known tramway expression 'notching up' to describe acceleration). The noise made by the star wheel and roller have an appeal to the interest of tramway enthusiasts as they give a realistic atmosphere of tramway operation.

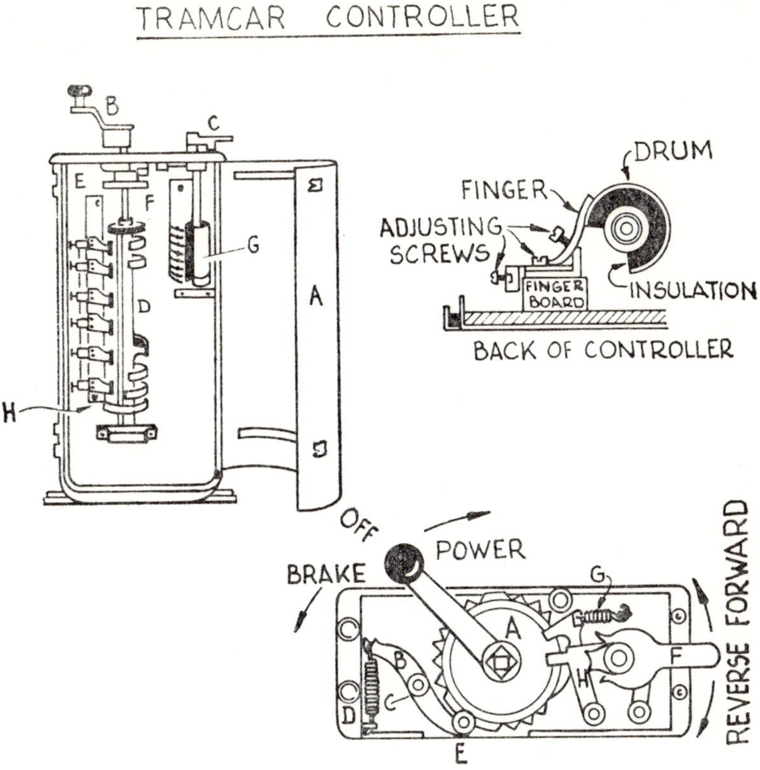

FIG 32 (above); FIG 33 (below)

The cover-plate of the controller has been removed in the illustration, Fig 33, and the controller handle and the key or reversing-handle are shown in the off positions. The star wheel is shown at *A* and many modellers have made this from a bicycle sprocket wheel. The lever *B* is pivoted at *C* and actuated by the spring *D* which holds the roller *E* in place and prevents the star wheel from moving in any but the notched positions. The controller key *F* in the off position prevents the power handle being operated by the hook *H* engaging with a notch on the star wheel. This hook *H* is also actuated by a spring *G*. When the controller key is moved to the forward or reverse position the 'V' shaped notches at the top of the reverse drum allow the pivoted hook *H* to be released from the star wheel by the spring *G* and the power handle moved. The controller key is detachable as when the motorman leaves the platform for any reason, after first securing the brakes, he always takes the controller key with him as the car is then immobilised. The controller power handle is also detachable but it is not very often detached from the controller. A number of modellers have very successfully constructed a controller handle from certain types of motor-cycle starting crank levers, mounting a door knob for the handle. The illustration, Fig 32, shows the method of constructing the cylinder viewed from the top. The controller back can be made from sheet metal but, in modelling it is normally constructed in wood as is the finger board. To the finger board are attached the fingers which, in modelling, are made from 1/2in by 1/8in copper strip cut to size and held in position by two adjusting screws, the screw shown on the left being the terminal screw for the wiring to the resistance. The springiness of the copper strip is quite sufficient, in a model controller, and there is no need to give one's self the additional work of installing powerful springs to operate each finger as in prototype practice as, again, the low voltage does not justify this. The contact segments on the drum which pass under and make electrical connection with the fingers when the drum is rotated can likewise be made from the same section copper strip. In modelling a controller, with the low voltage still in mind, many modellers have used a section of asbestos drainpipe for the drum, drilling it carefully at staggered intervals to take the contact segments. These contact segments have, of course, to be drilled in two places and the holding screws *countersunk* to provide a smooth surface for contact with the fingers, otherwise the fingers will bend. It is advisable to smear a little Vaseline on the fingers occasionally.

The principle of a resistance is as shown in Fig 34 and it is constructed on a wooden frame with an asbestos or ebonite base into which the brass holders are fixed by drilling and inserting brass nuts and bolts. Electric fire elements are very suitable for resistance wire and the appropriate lengths will have to be worked out by the experience gained in running the cars. The resistance wires should be firmly attached to the brass holders by screwing tightly; *solder should not be used.* The appropriate connections numbered 1 to 5 should be wired to the relative terminals of the controller fingers. Although these resistances were sited elsewhere on prototype tramcars, in modelling there is no reason why the entire resistance should not be installed inside the controller itself for compactness, underneath the reverse drum in the space occupied in

FIG 34 Principle of a simple resistance

actual practice by the blow-out magnet which is not necessary in scale modelling for small voltages.

Reverse switches seem to cause modellers some trouble but the principle is really quite simple. Good quality reverse switches can

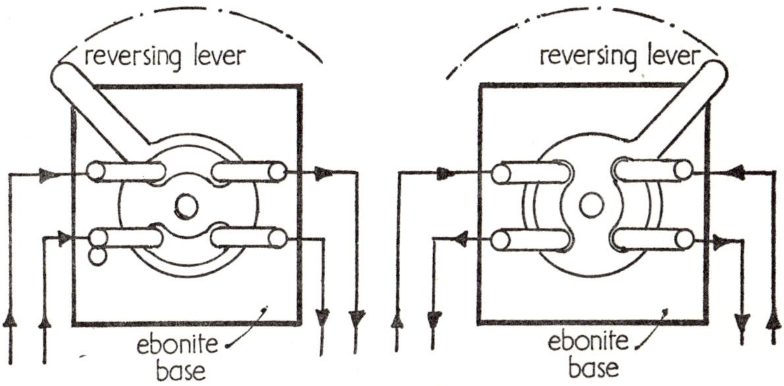

FIG 35 Principle of a reversing switch

also be purchased from ex-government surplus stores but, in case the modeller wants to make them himself, Fig 35 shows the quite straightforward principle upon which they work. The left-hand switch shows the connections in the forward position and the right-hand one shows the connections when reversed. By following the arrows it is quite easy to discern the changed path of the current, when the reversing lever is turned. Here again 1/2in by 1/8in copper strip should be used in constructing the segments and those mounted directly on to the ebonite revolving base (which rotates by the action of the lever) should have countersunk screws to fix them. If the four copper terminal segments are securely screwed down at the ends, it will be found that there is sufficient springiness in them to form good connection when the arm is turned. A spring washer should be inserted between the revolving ebonite disc and the ebonite base for ease of movement. If the modeller intends to construct his own exact replica of a standard type controller of the exact size used in actual tramway practice then the contacts shown in the illustration on a flat base are simply installed round the revolving reverse drum. In modelling, again due to the low voltage, this reverse drum can also be made from a small section of asbestos drain piping which can be drilled for fixing the necessary contact segments.

Whether the modeller builds his own controllers or purchases ready-made ones, two will be required to operate his model tramway, one for the 'up' road and one for the 'down' road. This applies whether the projected layout is single track and loops or whether it consists of long stretches of double track. It necessarily follows, therefore, that to emulate prototype operation in starting and stopping the model tramcars at selected places on the tramway, as every movement is governed by remote-control, sectionalisation is necessary, in addition to controller operation, if a number of tramcars are to be run. Now a model tramway differs electrically in its operation from its prototype as, while in the prototype tramways the overhead trolley wire is always the positive feeder and the rails the negative return circuit, in a model tramway they are used alternately due to the remote control. When, therefore, a model tramcar is reversed from remote control, the rails become the positive feeder and the overhead trolley wire the negative return. As the circuit is only 24 volts this does not matter in model tramway operation. But while one model tramcar on each independent feeder circuit may be controlled effectively from each controller (as the trolley wires are insulated from each other) when two cars are proceeding in the same direction they will be operated by one controller only. Assuming that the first car has reached the end of the double-line track and that it is necessary to hold it stationary to enable an oncoming car on the single-line to enter the double-line track in order that the cars may pass, while it is still desired to control the second following car from the controller effective on that circuit, the need for sectionalisation will be apparent.

There are two methods of electrically sectionalising a model tramway layout and both methods have their fervent advocates. Both are equally effective. The one method is by cutting the overhead trolley-wire at strategic points and inserting a 'section-breaker' as described in Chapter 10 to isolate a section electrically and leaving the track intact. The other method is by cutting *both rails* of the single track to be isolated and leaving a 1/8in gap in which can be inserted a small strip of Bakelite for insulation purposes and also in order not to impede the smoothing running of the car wheels, leaving the overhead trolley-wire intact. It will be found that the trolley-wire section-breakers are better for indoor layouts while the interruption of the circuit by cutting the rails is preferable for outdoor layouts. This is because the less strain on the overhead trolley wires on an outdoor layout the less likelihood is there of excessive repairs and adjustments, for an outdoor tramway has to stand up to more rigorous conditions and robust trolley wire overhead construction is essential.

Both methods of isolating sections are identical in principle and

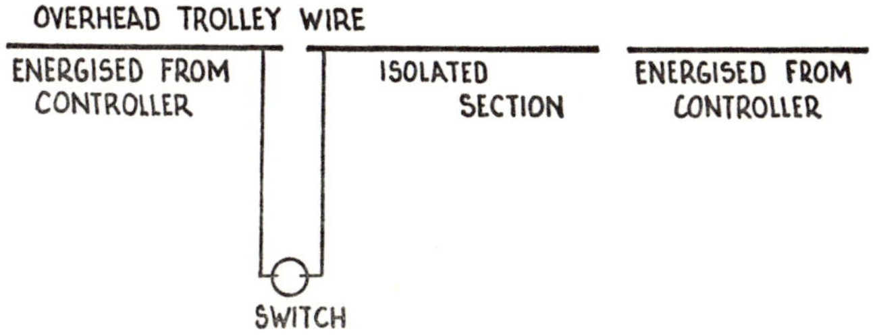

FIG 36 Method of sectionalising overhead

are quite simple. The illustration, Fig 36, shows the method of isolating the overhead trolley wire by a section-breaker, or two section-breakers if the isolated section is to be of any length (for instance, to hold several cars stationary). The feeder cables at either side of the section-breaker are taken back to the remote-control and are fed into either side of an ordinary house-type tumbler switch. The switch is just turned off if it is intended to hold a car in this section but if the operator wants the car to proceed normally he merely leaves the switch in the on position. Exactly the same applies to the rail method of sectionalising (Fig 37) only, in this case, the rails are securely cross-bonded at each

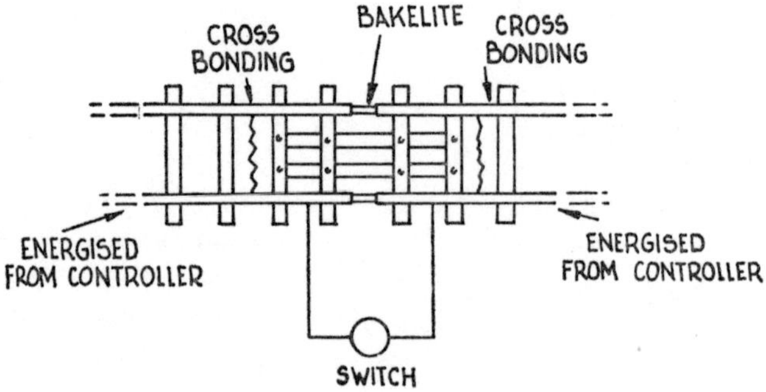

FIG 37 Method of sectionalising track

end of the isolated section for the sake of electrical continuity in both rails. The Bakelite or similar insulating strip can be shaped at rail-tread level and if constructed in the form of an 'L' shape, it can be screwed to the baseboard to keep it in place. In order to ensure that the rail-gauge is kept rigid at the point where the break is made it is advisable to insert two extra long sleepers underneath parallel to the rails and screw them firmly in place to prevent any side movement of the running rails. Whichever method of sectionalisation is used the wiring will be continuous on one side. If the rails are sectionalised in places then the trolley-wire will remain intact and if the opposite method is used and the trolley-wire is interrupted at the sections it will be the rails which will have the continuous circuit.

The illustration, Fig 38, shows the quite straightforward method of wiring up the tramway circuit. Starting from the left the connection from the house supply via the public mains is taken into the transformer and rectifier. The *low-tension* output will be clearly marked on this and the 24 volt circuit is commenced from this point, passing through an adequate fuse or an automatic circuit-

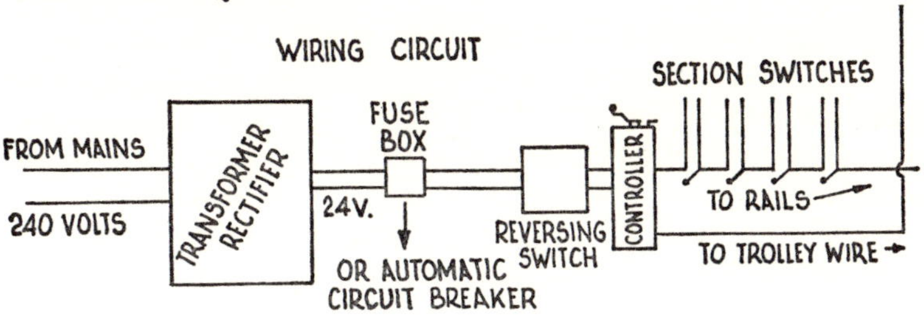

FIG 38 Wiring and sectionalising circuit

breaker to the reversing-switch. Although the reversing switch is shown independently this is only to emphasise that the 24 volt feeders *must* be wired through the reverser *before* reaching the controller (the switch itself can be incorporated in the controller, as previously described, provided it is wired as stated). The cables are connected to the two leads of the controller and, on the output side, in the illustration, one lead is taken directly to the trolley-wire while the other lead is taken to a control panel consisting of a bank of ordinary tumbler switches (or similar) which control the sectionalisation of the line. The diagram shows the switches in the 'open' position so that a car proceeding from, say, right to left along the route would stop immediately it reached the first section. By closing the switch and notching up again on the controller the car would proceed to the next section, when the process would be repeated, and so on. If, however, the operator wished the car to run through all the sections without stopping he would close all the section switches. Although the 'rail-break' method has been described, exactly the same principle applies to the 'trolley-wire break' system, in this latter case the trolley-wire being broken instead of the rails. The whole circuit from mains to model layout has been shown in simple form so as not to confuse the modeller.

In wiring up a model tramway system it is worth taking the extra trouble to make sure that all terminals are clean, the wiring entering the terminals is likewise clean and that both are securely tightened at the connections. If the connections are soldered, it is wise to clean both ends of the cables, twist together with pliers and then solder securely. Good sound electrical connections in the first place will obviate time needlessly spent in tracing faults afterwards. All cable used must be of best quality and to the highest standards of insulation.

Having wired up the system and tested it by running a model tramcar it may be discovered that there is a definite slowing down of the car at some point of the system, especially if it is an outdoor layout. Ruling out the possibility of the car itself being faulty (because it will have been 'bench-tested' at the completion of its construction) the slowing down may be due to two possible causes, the first being a loose connection. If the car slows down and then, suddenly, when passing a certain section of the track, speeds up for no apparent reason, it is obvious that the rail-bonding has become detached. This can be discovered quickly at night as there will be a small arc at the rail-joint when the car approaches it, or after the car has passed, depending on the circuit. The detached rail-bond will have to be re-cleaned and re-soldered on to the rail. If this is not the cause of the trouble the trolley-pole of the car (or cars) should be hooked down to immobilise it, the controller turned full on and a check made with a portable voltmeter at various points of the track with one lead of the voltmeter on the rail and the other lead on the trolley-wire, to see if there is a 'line-drop' in voltage anywhere—usually this will be found on the furthest point of the system away from the main feeders into the tramway. If no voltage drop is discovered, the same test should be made with a tramcar running. By getting an assistant to operate the controller and testing at various points with the voltmeter while the car is running, a voltage drop will quickly be discovered, if this is the trouble, at the point where the car is prone to slow down. Here again, some knowledge of actual prototype tramway practice will help. Unlike alternating current, direct current is prone to a 'voltage drop' and, to overcome this, in actual tramway practice the trolley wires were divided into half-mile sections, each section receiving an independent feed from the power station. The same principle is applied to model tramways, as a 24 volt line will sometimes show an appreciable voltage drop on an extensive garden layout. If this is discovered to be the cause of cars slowing down, it will be necessary to carry substantial-gauge insulated copper cables underneath the baseboard of the track from the main feeders of the controller and solder them, at intervals, where there is a 'line-drop' in voltage, to the trolley-wires. In addition to the inadequate supply of current at the furthest point of the system it sometimes happens that a model tramcar will slow up at a slight gradient in the line or when rounding a particularly sharp curve. If this is so, an extra 'boost' feeder should be installed at these points. Before installing extra booster cables it is advisable to make quite sure that at the section where the cars have been slowing down the rails and overhead trolley wires are clean, as dirt or oxide may be the cause of the loss of power.

13 PAINTING, LINING AND LETTERING

It is a truism often heard that 'many a splendid model is spoiled by shoddy painting and lettering' and going round model exhibitions one finds several admirable models, whether railed rolling stock, ships, or aircraft, where the modeller has given considerable time and trouble to constructing the model, the detail of which is perfect, but has completely spoiled the effect by being in too much of a hurry with the vital task of painting his model. This is a great pity because, with the present modern range of paints and lettering and lining sets available especially for modellers, there is absolutely no need for any model to be spoiled in the last stage—the painting stage.

Any timber used in making a model tramcar *must* be glass-papered *before* use as difficulty will be experienced in getting to the smaller and more inaccessible parts afterwards. Although timber sold in model shops appears to be quite smooth a close look at it under a watchmaker's eye-glass will disclose that this is not so. Left in this state, without being glass-papered, inequalities in the surface will show up after the model is painted. Quite considerable sections of a model tramcar have to be painted or varnished *before* being inserted into the model. The interior panelling, the ceilings, the seats, the staircases, the flooring, underneath the canopies and the inside of the dashes. This latter is frequently forgotten and it is practically impossible to paint it properly when the controller and staircases have been installed. The golden rule is, immediately an interior section has been finished, to think 'when installed, will this be inaccessible?' Another golden rule is never to be in a hurry when painting. Paint-drying cannot be hurried and it is well worth a little patience to leave undercoats and final coats plenty of time to dry and harden. If a slow drying paint is used, either for the undercoat or for the final coat, it is of paramount importance that the model is painted and left to dry in as dust-proof a room as possible.

The type of paint and varnish for painting a model tramcar needs the greatest care in selection. Normal paints and varnishes are far too coarse for the immaculate finish required in modelling and, although it is possible to use an ordinary priming-coat first, it will have to be mixed with 'thinners' before use. The most satisfactory paint, which gives a model a superb finish, is the modeller's paint manufactured by Humbrol, obtainable from all model shops. It drys in two to four hours and gives a high gloss finish on wood, glass, metals, plastics, and hardboard. Ideal for the modeller, it can be obtained in the very small quantities necessary, in a wide variety of colours. A small selection of artist's brushes is required, of soft texture, not forgetting to obtain two or three flat ones also.

It is essential to clear all dust from the model before starting to paint, all glass-paper dust, tiny bits of glass or Perspex, and dust from the atmosphere; attention should particularly be given to corners as these particles tend to accumulate there. A fine camel-hair brush should be kept for this purpose. The primer can then be applied and left to dry thoroughly, followed by the undercoat which is also left to dry completely. A mistake often made by amateur painters is to use too much paint which causes 'running' of the paint and it is particularly advisable to watch for this 'running' to collect in corners of the model. On completion of the undercoat, after it is dry, it must be *lightly* rubbed down with very fine glass-paper to ensure an absolutely smooth surface for the next undercoat which is treated similarly. The model is now ready for the final colour.

As tramcars have a certain amount of lettering, numerals, and crests it is preferable to use a 'flat' finish for the sound reason that after these are applied, the surface can be given a coat of thin varnish. Here again this flat finish must be very smoothly applied sparingly in order that 'running' of the paint does not occur. It is a good idea to have a 'test piece' each of wood and metal to try the flat paint on, first applying the thin coat of varnish when this is dry as varnish tends slightly to alter the shade of the paint. It is only by trial and error on this 'test piece' that a faithful reproduction of the livery used in the prototype tramcar can be gauged.

The exteriors of prototype tramcars made use of a number of colours, in some cases the waist-panels and dashes may have been of green or red while the rocker panels were of cream, grey, or white while the roofs (of single-deck cars and of covered-top cars) may have been either battleship grey or white. The modeller should not attempt to paint another section of the car until one colour is absolutely dry. If the model is in two tones, it is advisable to paint the rocker-panels, window pillars, and lintels first as these will be in the lighter colour. After these are quite dry attention can be turned to the waist panels and dashes. Where the colours meet it will be necessary to cut a piece of painter's 'masking-strip' which can be obtained from any painter or signwriter as only a very small quantity is required. This 'masking-strip' should be cut to the size required and placed over the section already painted before the second colour is applied. When this latter is dry, the 'masking-strip' is easily removed disclosing a very clean line between the two colours. Normally these two colours are divided by a clear thin black line. It is extremely difficult to insert this line either with a very fine brush or with a pen and recourse should be had to 'Letraset' lining-paper which is ideal for this purpose. By following the directions carefully excellent definite straight dividing lines will result.

One aspect of giving a model tramcar a professional finish which has caused modellers no little difficulty in the past has now been made extremely easy by the advent of 'Letraset', waterslide type transfers, and goldsize type transfers. Most stationers stock 'Letraset' and it is advisable to use it for side destination-boards, rocker-panel or lower panel wording (METROPOLITAN ELECTRIC TRAMWAYS LTD in 1/8in lettering on the bottom left-hand corner and CHRISTOPHER JOHN SPENCER General Manager on the bottom right-hand corner, for example). In the example given the lettering is in black. On the sides of the stairs the words 'WAIT UNTIL THE CAR STOPS' were sometimes painted in gold paint and, here again, 'Letraset' or the waterslide type or goldsize type transfers can be used. These two latter types should always be used for the numerals of the car fleet number or (if they were used in the prototype being modelled) the wording on the rocker panel such as READING CORPORATION TRAMWAYS.

The method of application differs for each of the three types of lining or lettering. It is first necessary to scale the actual size of the lettering required and, in the case of 'Letraset' ask the stationer for his demonstration book of the various sizes and types available. This is quite a comprehensive book as there are very many types and sizes and also differing colours. Having ascertained the correct 'Letraset' style and size of lettering, its application to the model requires a little care. In the case of lining, whatever colour and width is used, it is cut from the sheet, the back paper removed, placed carefully in position and then rubbed on with the rounded edge of a ball-point pen. The thin tissue 'holding-paper' will then easily be removed leaving a perfect line in position. Lettering is done in exactly the same manner, each individual letter to make up the necessary word being rubbed down separately. In order to ensure that the letters are in a perfectly straight line a length of cotton should be fixed down with a strip of Sellotape at each end to guide the eye. The lining or lettering should be left for an hour or so to settle and then a coat of clear varnish applied *over the whole extent of the panel* and not just over the lettering as, if it is applied only over the lettering the varnish will show up to bad effect afterwards.

Waterslide type transfers are applied quite differently. These can usually only be purchased at model shops and it will be necessary to search through a selection of sheets to find the correct scale types. Again the letters have to be cut out and applied by damping the holding paper with water and *sliding* the relative letters into position, the thin tissue paper on top being removed carefully when the letter is in position. The water should be used sparingly otherwise letters already in place will be displaced.

Goldsize type transfers differ from the foregoing as they are in reverse when view from the front, the holding-paper being thicker. They are normally used for rocker-panel or waist-panel wording as in the words LONDON TRANSPORT or the car fleet numbers. They are applied by first cutting out the complete words or numbers to be used, placing them in correct order and then brushing a light film of methylated spirit over the area where they are to be applied. As methylated spirit evaporates quickly, the reason for lining up the letters or numerals beforehand will be apparent as they have to be positioned rather quickly. They should then be left for one or two hours to harden before the holding-paper is soaked off gently, leaving the gold lettering intact. If the goldsize type transfers are unlined, the black outer lining can be applied by cutting strips of 'Letraset' to encase them, applying this as previously described. Whichever of the three types are used, after application they should be left for some time to set before the final coat of clear varnish is applied and it needs emphasising again that this should be over the whole area and not merely over the lettering, or lining.

Although the foregoing deals with the normal type of lining and lettering which has been simplified in recent years there are two aspects of craftsmanship which still cause a certain amount of difficulty as they are intricate. These are the ornate scrollwork at the corners of the linings, if the model is to be a replica of an early type tramcar, and the coat of arms on the waist-panel. While a variety of scrollwork is obtainable, by diligent searching, in the three methods described, it may not be in the exact styling required. In a number of cases, with careful cutting and application it is possible to adapt it but both with the corner scrolling and the coats of arms meticulous hand-painting is the only solution. Before attempting the coat of arms some research is essential to copy it exactly. While the exact outline and shape of it can be copied from a normal black and white photograph, it is advisable to draw it from this to exact size, inserting the lettering at the same time. It will then be necessary to seek the co-operation of the corporation or local authority whose coat of arms is to be copied. Usually the archivists are very helpful when the need is explained and information is forthcoming. In some cases concerning the earlier types of tramcars an old brochure concerning some past event in the city has been unearthed on which was a coloured coat of arms, in other cases the same coat of arms is still used on corporation transport at present in use. It is well to verify the styling, however, as it may have been altered over the years as a number of researchers have already discovered. In the case of company-owned tramcars research will have to be undertaken with the Tramway & Light Railway Society or the Tramway Museum Society as they own copies of some of the company-owned systems' coats of arms. When the colours, gold lettering, and insignia of the coat of arms have been correctly ascertained, the previously prepared full-size drawing should be correctly completed by painting and lettering in the proper colours. It is only by this method that the smaller replica can be copied on to the sides of the model. Before attempting it on the model it is better to prepare a 'trial side' of the tram-

Page 143 (above) Mr Robert Whetstone's 1½in to the ft scale model of Birmingham Corporation bogie double-deck tramcar No 700, showing his method of assembly of the upper and lower saloons; *(below)* Liverpool Corporation 'cabin-type' tramcar No 843 constructed in 3/4in to the ft scale by Mr Leo Taylor, a striking and well detailed model expertly painted in the impressive olive green livery of the Liverpool fleet

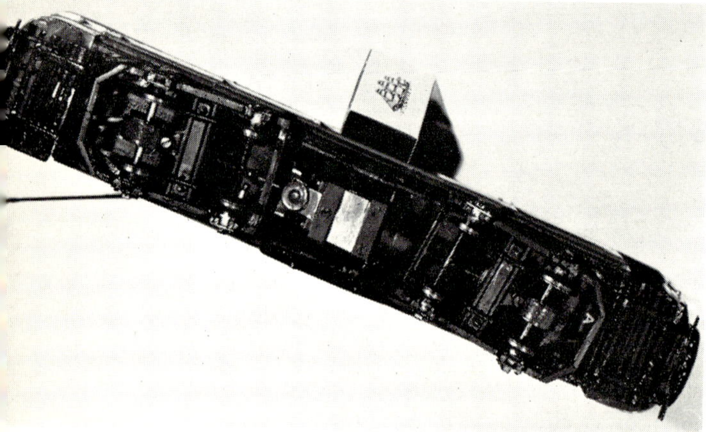

Page 144 (top) Halifax Corporation single-deck four-wheel tramcar No 105, constructed to 3/8in to the ft scale by Mr Eric Thornton; it runs, with the rest of his fleet of models representing many systems, on his extensive layout at Sherborne. The prolific detail attained in this small scale deserves close study by modellers; *(centre)* Llandudno single-deck bogie car constructed in 3/8in scale by Mr Eric Thornton, showing the special motor mounting and flexible drive adopted in this scale; *(right)* Blackpool railcoach No 319 absorbing the attention of a small boy on emerging from the subway on the garden model tramway at Glastonbury

car and have several attempts to test one's ability and to gain experience in so intricate a venture. The operation will be made easier by cutting out the outline, to scale, of the shield or crest to 1/32in under size and carefully scribing round it to get the exact outline, the 1/32in being to allow for the scribe-mark. The tiny area available for the various coloured paints can only be treated with the finest camel-hair brush obtainable and, even so, it will be necessary to trim it down even finer. The same brush can be used for the gold paint but, if this is not successful in scribing the tiny letters, a mapping-pen should be tried. Some modellers have greater success with one than with the other. Before attempting to use gold paint (and this should be bought from a model shop in a small quantity) the varnish must be poured off to leave the gold at the bottom of the bottle. A minute quantity must be mixed with an equally minute quantity of turpentine to make it flow either through the brush or through the mapping-pen with whichever one is most proficient. The overriding principle in using gold paint in this manner is to control the flow so that it is even and does not congeal in blobs at the end of the brush or pen. Described in this manner it gives the impression of being a painstaking task. And so it is but it is well worth while to contemplate the result and admire an impressive tiny coat of arms on the side of the model which so greatly enhances its whole appearance.

If the gold lining has been applied to the model by means of one of the proprietary lining sheets, the ornate scroll work at the corners can be completed by the same method as the one described for the coat of arms. In both cases it is important to remember that some ingenuity will be necessary in maintaining a steady hand. Some modellers rig up a wooden frame in which they can insert the model, using the frame to rest their painting-hand on, others rely on what is colloquially known as a 'sign-writer's mop' which is easily made from an 18in by 1/2in strip of wood with a rounded ball-shaped piece of cloth securely attached to one end.

Other types of paints are, of course, available but the correct modeller's paints are by far the best. Cellulose paints are not suitable for models unless the modeller is very expert in the use of these. Occasionally some modellers have been successful in obtaining from a local authority the exact paint used in their present-day vehicles if it is the same as formerly used on their tramcars but it has been found unsuitable for model-work, unless again the modeller is very expert at thinning or altering the texture of the paint.

In addition to painting the actual coachwork, the ancillary equipment of a tramcar has to be painted the correct colours of the prototype being copied. This, of course, differed with the different systems and can only be ascertained by research. Stanchions were usually painted black and where they incorporated a handrail this was polished brass. Controllers were painted black but the tops were brass and, in some cases in latter years, chromium. Single-trucks and bogies were of differing colours depending on the system's choice; red oxide or grey was popular on some tramways, providing a sharp contrast with the livery of the body of the car but other systems preferred to paint them the same colour as the predominant colour of the car. Bristol, for example, matched their cars' trucks with the well-known blue of their fleet. Lifeguards were normally painted the same colour as the trucks. Iron scroll-work and protective meshwork (on open-top and open balcony cars) was painted black and black was also used on beading edges and collision fenders. The inside of the dashes and the inside of the upper deck panelling (on open-top cars) was a choice between the usual light or dark grey or, in some cases, a medium shade of brown, or black. The outside of the upper deck panelling could be either the same colour as the rocker panels or the predominant colour of the car. The undersides of the canopies were mainly cream, the window pillars being coloured likewise, although the 'fluting' on the corner window pillars was occasionally of a different contrasting colour (for example, Bristol's 'fluting' was red). Floors, floor-slats, stair-treads, and risers were usually dark grey or dark brown. The interior of lower and upper saloons was finished in a pleasing shade of light oak, with the door panels occasionally being in a slightly deeper shade of light oak. In early days the interior ceilings of both lower and upper saloons were either in 3in strips of polished longitudinal wood slightly grooved at the edges, alternate strips being light oak contrasting with similar alternate strips in dark oak, or the entire ceilings were in plywood with a light brown finish and a decorative surround of a deeper brown scroll work. Latterly most tramcar ceilings were painted white.

Naturally only a general idea of the painting scheme is given as it differed with the different systems, each being distinctive, but the general principles described were adhered to in the main. As in other aspects of building a scale model tramcar, the modeller will have to undertake the necessary research to ascertain the correct livery of the car he is building. There are a number of avenues open to him to do this. The Tramway Museum Society have over forty tramcars fully operative and running on their Museum tracks at Crich near Matlock which have been painstakingly restored and the accuracy of their liveries is undoubted. A number of museums throughout the world have preserved tramcars or, at least, one tramcar representative of their systems. Some museums had models of their tramcars professionally made, before their local tramways were abandoned and there are many such excellent models, in the correct detailed liveries, on public display. It is advisable for the modeller, when he has located a prototype, to take a spare copy of his scale drawing to the museum and record the correct colours on it. Quite a number of professionally produced scale drawings do give the correct colours needed. While engaged on this research, the modeller should also take detailed notes of the wording on the prototype (for example, in all cases the name of the general manager on the lower corner of the rocker panel and, in some cases, the class or type of car was also painted on the opposite corner). It is this keen attention to detail which brings out the character of the finished model—and nowhere is this more apparent than in the detailed painting finish.

One striking aspect of a model tramcar which stands out above all the rest at first glance is the destination indicators. These take the form of destination boxes on the front of the car and sometimes on some systems at the sides while, in older-type cars they take the form of destination boards, these latter usually giving intermediate towns or streets and the boards being interchangeable according to routes. Destination blinds can be made to revolve on rollers, as in actual practice, by means of bevel-gearing or just straightforward rollers but a number of modellers content themselves with a static blind which can be made to slide out, from the side, to change the destination indicated. Destination blinds in Great Britain are white lettering on a black background but continental blinds are, in the majority of cases, black lettering on a white background. The continental type present no problem as the appropriate size black lettering is available in 'Letraset'. Previously the British type of destination blinds were tedious to make as the block capitals were left white with the blacking-in done round them in indian ink. Fortunately a simpler method has now been discovered. In both cases the blinds have to be of draughtsman's linen as ordinary paper will curl up in the heat of the destination lighting bulbs behind. To obtain white lettering on a black background the 'Letraset' lettering is put down *but it is not burnished down.* The whole of the linen is then painted over with indian ink or drawing ink several times, allowing each coat to dry before applying the next, to get the correct degree of blackness in order that the illumination behind the blind, at night, only shines through the actual lettering. When the completed frames are black enough and quite dry, the 'Letraset' is lifted with a strip of Sellotape, leaving the letters correctly reversed. Route numbers can be undertaken in exactly the same manner.

Side destination boards, after being cut to scale size, are given a primer of white paint and a final coat of Humbrol white enamel. When this has set thoroughly, a length of black cotton is then fastened across it to ensure that the letters are correctly and evenly lined up. With patience and care black 'Letraset' letters of the appropriate scale in block capitals are then applied to form the route words. These side destination boards used to be very difficult to do but 'Letraset' has completely simplified the process. A coat of clear varnish will finish off the boards to professional standards.

It may be that the modeller having successfully constructed his first model tramcar does not wish to confine his interest to just that one and has in mind additions to his fleet. A number of tramway modellers build several scale models of different type cars of the same tramway system while the majority model different cars of a number of systems. Whichever inclination the modeller has the painting of the lineside and 'street' equipment requires consideration. Here again there is no hard and fast rule. Although most tramway systems painted their trolley standards in a sober shade of green, some adopted dark brown (in the Black Country) and there were some examples of aluminium painting, and also black. The 'STOP' signs varied with the systems and, to the modeller, it is necessary to scale them down to the correct scale. Some were mounted on the trolley standards and others had their own standards. The wording 'ALL CARS STOP HERE' (Fig 39) or 'CARS STOP BY REQUEST' (Fig 40) can be neatly painted in

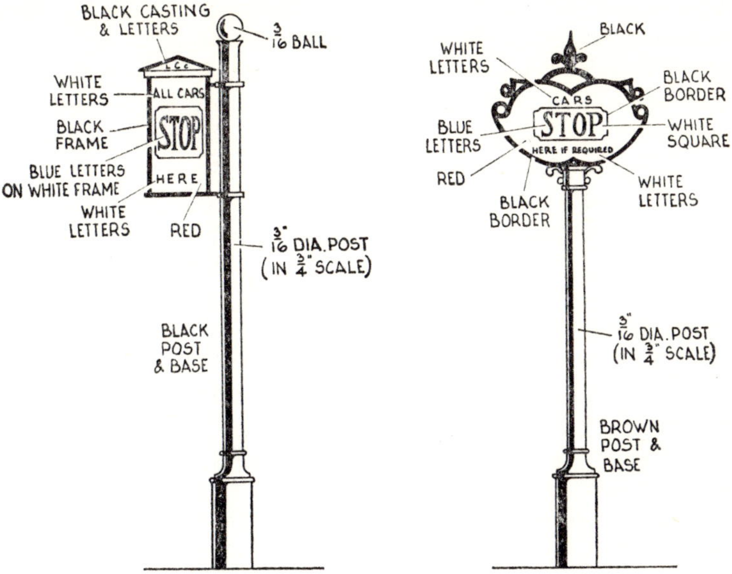

FIG 39 *(left)* London County Council obligatory stop; FIG 40 *(right)* London County Council request stop. Showing distinctive painting for each

white on a black background or black on a red background, or white on a red background according to the choice of line one is modelling. The tramway waiting rooms are quite easy to construct as they were very similar to the present street waiting rooms and also painted similarly, usually in medium green.

The painting of the depot 'roads' was usually a white numeral in a black circle, '1', '2', '3', '4' and so on. The depot offices also provide scope for the modeller's ingenuity: 'STORES', 'TRAFFIC OFFICE', 'SUPERINTENDENT' and similar designations, again normally painted in white on a black background. Additionally each trolley standard bore a number, methods of numbering differing as some systems numbered '1' from the depot, if it was situated at one end of the system while others numbered their '1' from a terminus. These tiny numbers are easily done in 'Letraset' remembering to put a coat of clear varnish on afterwards. On all tramway systems were a number of compulsory stops known as 'Board of Trade Stops' usually at half-mile intervals or at the top of sharp gradients or near traffic junctions. The object of these compulsory stops was mainly to ensure that the motorman was able to stop the car. Had he been taken ill, for instance, and fallen down on the platform leaving the power on, the conductor would immediately stop the car from the rear platform. These 'Board of Trade Stops' were clearly marked by about three feet of the appropriate trolley standard being painted pillar-box red circling the pole about a third of the distance from the top. If there were two standards supporting a span-wire on each side of the road, both were identically painted red.

Another adjunct adding character to a model tramway layout concerns the judicious use of scale advertisements on the walls outside the depot. These are easily obtained from a variety of labels on household wrappings, tins, and bottles again remembering, if the layout is outdoors, to ensure that a light coat of clear varish is brushed over them after fixing. Street signs, 'NO WAITING' 'WAITING RESTRICTED' and similar types are made by neatly cutting out from booklets issued by the motoring organisations or the Ministry of Transport and they are already coloured. By use of the imagination the modeller has a wide scope for making his layout an exact replica of the original.

14 ELECTRICAL CIRCUITS FOR POWERING AND LIGHTING THE MODEL

Unlike the prototype tramcar being modelled, which contains a very complicated and comprehensive wiring system, the actual model tramcar due to its being remote controlled has a comparatively straightforward traction wiring circuit. The type of cable used is ordinary house-wiring cable capable of taking 5 amperes and different coloured cable should be used for the different circuits, to assist in tracing any faults which may develop at any future date. Again, provided the original wiring is sound, it is only very occasionally that a fault will develop but, on these remote occasions, unless different coloured wiring has been used, considerable waste of time will ensue in trying to trace it.

The two circuits shown in the illustrations are the most commonly used. The simple parallel wiring is for two 24 volt motors wired in parallel (Fig 41) and the series connections (Fig 42) are for the benefit of the modeller who has not been able to procure 24 volt motors but has managed to get 12 volt ones, as sometimes happens. In each case a reverser switch has been shown in the circuit. This is additional to the remote-controlled reverser on the controller. The object of an additional reverser on the models is to facilitate easier working of the model layout as, when the fleet is extended, it will be appreciated that individual reversing-switches on the models will be a great advantage. In all cases, if the models are running on tracks at exhibitions, each car has to be fitted with a separate reversing switch as, otherwise, if the operator attempted to reverse one car from his remote control, all the cars on the track would likewise reverse, causing chaos. These

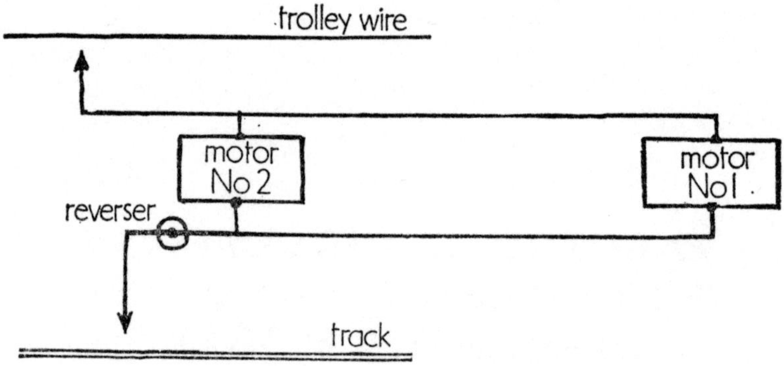

Fig 41 Traction circuit, parallel connections

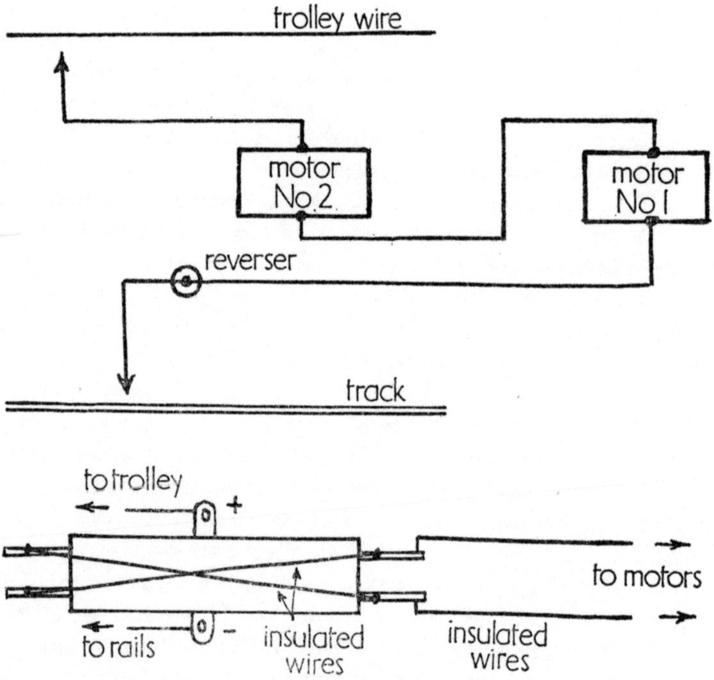

Fig 42 Traction circuit, series connections; (below) method of wiring DPDT toggle switch. This type of switch can be used in the main layout and also on the individual cars

small reversing switches are usually mounted on the models underneath one of the platforms within easy access of the operators. Now the modeller can easily make a small reversing switch for this purpose on the lines of the one described in Chapter 12 but there is now available a much simpler and neater product which can be obtained from any radio spares dealer. A certain amount of care is required in asking for the correct type as two types of switch look identical but it is only the *double-pole* switch which will reverse the model tramcar. It is known as a DPDT Toggle Switch and the modeller will have to make a small 'L' shaped bracket for the purpose of mounting this switch underneath one of the platforms of his model, with the switch-arm pointing sideways and sufficiently recessed not to protrude over the side of the car. The 'L' shaped bracket (easily cut from aluminium sheet) is screwed on to the underside of the platform, a large hole being drilled to receive the switch which is provided with a 'keeper' holding nut for fixing firmly. The reversing switch is wired into the main traction circuit as shown. All connections to motor and reverser terminals and all jointing of cables *must* be securely soldered; it is also important that any cable from which the insulation has been stripped (to solder on at terminals or joint at any point) *must* be taped with insulating tape as any bare cable can, of course, cause short circuiting and, in the confined space into which the wiring of a model tramcar is restricted, electrical faults are sometimes time consuming to discover.

The lighting circuit of a tramcar is best understood if the principles of actual prototype practice are first understood. As the line voltage of tramway systems is 550 dc the lighting is connected in circuits or 'banks' of five lamps of 110 volts each, wired in series. Thus in a model tramcar this can successfully be imitated by circuits of five 6 volt bulbs, wired in series, making 30 volts or actually 6 volts above the model line voltage of 24 volts. It has been found in practice that the 'spare' 6 volts acts as a safeguard to prevent the bulbs burning out should there be any slight surge in the line voltage as sometimes occurs. Additionally the bulbs last much longer by not being fed to full capacity.

By law tramcars were required to show a white dash or headlight and a red rear light—in later years two of each. The illustration, Fig 43, gives the wiring diagram for lighting a single-deck car with 15 lamps. It will be noticed that 17 lamps are shown but

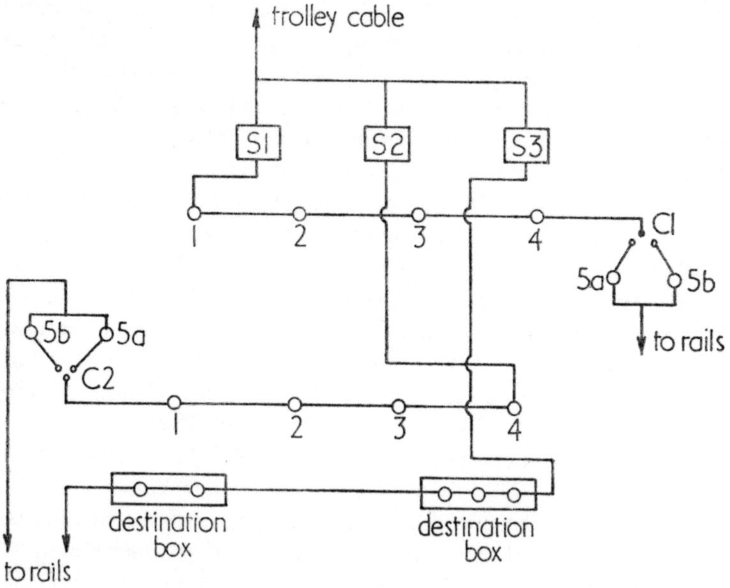

FIG 43 Tramcar lighting circuit wired in series (with three-point reversible switches for head and tail lights)

as one headlamp and one tail lamp are not in use at the same time, the 15 lamp circuit matches up to three circuits of 5 lamps each circuit. The main lighting feeder is taken from the trolley cable and distributed to the three switches S1, S2, and S3. The switch S1 controls the interior lights 1, 2, 3, and 4, the fifth lamp being the rear red light (5a) or the dash lamp (5b) according to the position of the three-point switch (C1). The circuit is completed by being connected to the car truck frame. Switch S2 operates a similar circuit 1, 2, 3, 4; and 5a or 5b on the opposite end of the car are controlled by another three-point switch C2. Switch S3 controls the lamps in the two destination boxes, three lamps in one and two lamps in the other. In the case of a top-covered car an additional circuit of lights on the foregoing principle would be installed. If, in modelling, independent switches are not desired for these individual circuits the whole lighting of the tramcar can be wired directly at 24 volts for each lamp *in parallel wiring* throughout the car. Small bulbs in this voltage can be obtained from any electrical components shop. The three-point control switch for the head and tail lights operates in exactly the same manner for this circuit. The type of switches used for controlling lights on model tramcars are the very small press switches of the type used in bedside lamps, one press to light the lamps, one press to extinguish them. These tiny switches can easily be sited unobtrusively underneath a platform.

When the trolley cable is installed down through the car to the underside of a platform, two circuits are immediately taken from it, one circuit for the traction current and the other circuit for the lighting current. Emphasis is again placed on the desirability of using different coloured wiring for each circuit.

There are, of course, a number of permutations in the wiring described which the modeller may wish to arrange to suit the lighting arrangements on his particular model. At least one modeller has installed flashing indicator lights on his models, actuated by contacts on the bogies which engage with copper sliding strips as the car rounds a bend bringing the flashing mechanism into circuit causing the indicator orange lights to flash realistically. Another modeller has wired the circuit for changing over head and tail lights into the changeover switch for the motors which, with the aid of two Mullard Type Serial BY.100 rectifiers automatically changes over the head and tail lights when the car is reversed, without the aid of a separate three-point switch.

Now it will be realised that the lighting circuits in model tramcars which have been described will only operate when the current is on the line. The disadvantage, in model tramcars, is that when the controller power is switched off, the lights of the model tramcars will extinguish. One modeller has overcome this by installing small batteries underneath the seats of his models but this is only possible when the lower saloon seating is longitudinal. It is not possible with small transverse seating. Furthermore, with the profusion of lights on a tramcar such batteries do not last long. The disadvantage of the lights going out when the model tramcar is stopped has been overcome by not switching the controller off entirely. By keeping just one 'notch' (the first and lower one) on, when the cars are lit up, sufficient current will pass to keep the cars illuminated, without moving the car. When this is done, a slight 'humming' will be heard from the motors but on no less an authority than that of the late Mr P. W. Lawson, the renowned superintendent of the Kyotts Lake Road tramcar repair works of Birmingham City Transport, who was once questioned on this point, no damage to the motors results from this practice. Mr Lawson also had his own model tramway layout in addition to being responsible for the high standard of maintenance of the entire large fleet of Birmingham tramcars.

Model tramcars are designed for remote control operation as it is not possible to make the tiny controllers on the models operative in small gauges. The cardinal point to remember is standardisation. If the modeller has decided on a certain type of motor, then he must keep to this type in any other models he constructs. The same applies to car lighting; having settled for series wiring and 6 volt bulbs he must likewise keep to this throughout—remembering that, if he adopts this type of circuit, he will have to have his street lighting in circuits of 5 lights and so on. It is, of course, possible to have some cars with two 12 volt motors in series and others with two 24 volt motors in parallel, provided they are the same type of motors but, beyond this, if a modeller has a variety of differing lighting circuits and differing bulbs in various cars he will cause himself endless trouble. The most satisfactory lighting system for modelling has been found to be individual 24 volt bulbs wired in parallel throughout as the filaments of these bulbs are more robust and give very little trouble.

There are two schools of thought amongst tramcar scale modellers regarding the provision on a scale model tramcar of a protective device in the car's electrical circuit. Some hold that as the models are remote controlled and the circuit from the power-unit to the controller is, itself, protected by fuses there is no need for an additional fuse in the car wiring circuit. Others regard a fuse in the car's wiring circuit as of paramount importance. This is a matter for the individual modeller to decide having regard to the characteristics of his layout. There have been a few rare occasions when upon investigation of smoke emerging from a model tramcar it has been discovered that a dead short circuit on the car has caused the wiring to ignite due entirely to the fuses at the power-unit being set too high. At the same time it has to be remembered that, as the system extends and more models are built, the capacity of the fuses of the power-unit will have to be high enough to pass sufficient current for operating several models, the street lights and other ancilliary equipment on the line. As a model tramcar only draws sufficient current for its own operation, it necessarily follows that the fuses on the model are set much lower than those which protect the output of the power unit. It is difficult to state what the exact capacity of the car fuses should be without knowing the power of the motors, the resistance being used and the weight of the car; only the individual modeller can discover this by trial and error on his particular layout.

One important aspect all too frequently disregarded in model tramcar wiring, probably because the modeller, having completed his model, is eager to run it on the track, is to ensure that all the car wiring, both the traction and the lighting circuits, is securely attached to the underneath side of the car, well away from the moving parts of the mechanism. Unattached wiring can easily rub against wheels, particularly gear-wheels, or become entangled with swivelling bogies, destroying the insulation of the wiring and causing electrical faults frequently difficult to discover. A little care in this respect is amply repaid by the eventual success of the model over a long period of time.

15 TRAMWAY MODELLING IN 3/8in TO THE FOOT SCALE

The ground covered thus far has been related to the larger scales but there are many modellers who, for a variety of reasons, find it convenient or necessary to build in the smaller scales. Each modeller will decide just exactly what he is looking for from his hobby.

Most of the larger scales are primarily designed to operate out of doors but, in the smaller scales, little is known of the behaviour of track and overhead equipment when subjected to the rigours of varying climates. It is therefore intended to concentrate on indoor layouts and also keep to 3/8in to the ft in general although every facet is equally good for 7mm scale. The main principles of tramcar modelling described throughout this book are applicable, although in a smaller scale considerable modification will be necessary having regard to the further restricted space of the small models, track, and overhead equipment. This is not to say that the detail in the smaller scales will not be equally impressive and satisfying: the illustration of Mr Eric Thornton's 10mm scale model of the single-deck Halifax Corporation's car No 105 demonstrates the excellent wealth of detail attainable. (Plate p 144) The high standard of truck construction, lifeguards, and the coachwork with the handrails, door-handles in addition to the exquisitely fine painting, lining, lettering, and the crest-work serve to show what can be achieved in this scale.

As most of the aspects of tramway modelling have been covered throughout this book, the descriptions of the smaller scales are limited to the divergences in so far as they will affect the different problems to be encountered and overcome in these scales. At the same time, many of the questions which will arise in smaller scales can be resolved by reference to the information already available in chapters applicable to the larger scales. All tramway modelling lends itself to the challenge of considerable ingenuity and nowhere is this more demonstrable than in the smaller scales.

If the track work is well laid and true to gauge then springing is not absolutely essential in small scale but if springing is adopted the riding qualities of the model will be that much better. Here again a model with four wheels without springing will keep to the track far better than railway rolling stock, due to the presence of a continuous check rail in the track, so that even if one wheel rides up onto the tread the opposite wheel and check rail prevents a derailment. The use of commercial coarse scale wheels is not always possible with a model tramcar due to the limited clearance between the front face of the wheel and the truck side frames but this problem is overcome if any good friend with a lathe can be persuaded to turn the wheels to something nearer to the correct scale widths. Dimensions for fine scale wheels and axles and for 3/8in scale are given in Fig 44; and these have been proven over

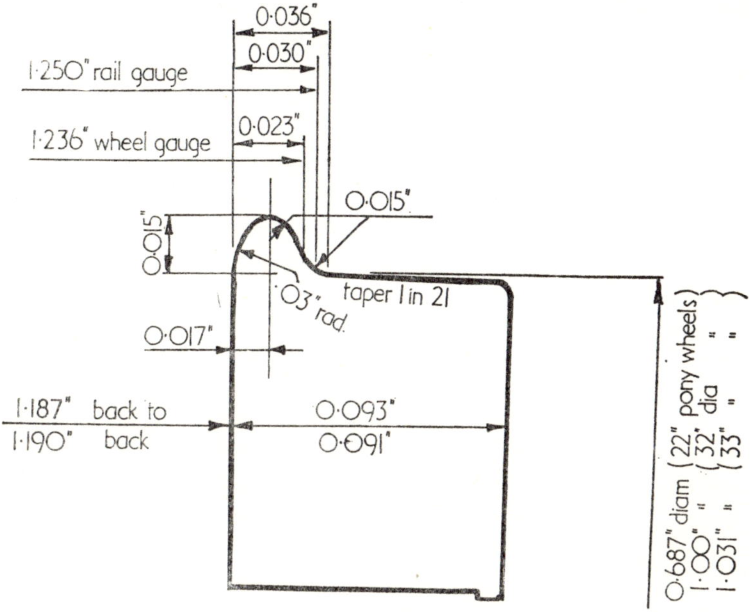

FIG 44 Fine scale tread and flange dimensions for 3/8in scale. The rail gauge, wheel gauge, and back-to-back dimensions shown are correct for gauge 0. For gauge 1 they will be 1·750in, 1·736in, and 1·690in respectively

many years. It should be emphasised that if fine scale dimensions are used then, because of the small flange depth, fully sprung trucks are an absolute must to ensure that wheels follow any undulation in trackwork.

The diameter of tramcar wheels varies only slightly. At the turn of the century driving wheels were invariably 30in diameter (15/16in) on the tread, but as steel tyred wheels began to replace the early chilled cast-iron ones the diameters began to increase first to $31\frac{3}{4}$ (1in nominal) over tread and in later years to 33in ($1\frac{1}{32}$in) diameter.

Most tramway wheel centres were of rather ornate and Victorian design, having six or eight curved spokes of a 'T' section and for standard gauge were dished inwards into the bargain. For 3ft 6in gauge wheels the spokes were still of 'T' section but were flat. For the modeller who intends to construct several trucks, simple jigs and gauges should be made and the wheels finished on their axles turned up as one unit in a lathe, thus ensuring that all wheels are identical in gauge and true in diameter. Since most old tramcars are remembered for the oscillating side movement, it may not perhaps be necessary for the wheels to run absolutely true if this effect is desired.

If two-rail operation is envisaged an insulating bush should be fitted to one wheel on each axle and care should be taken to assemble both insulated wheels on the same side of the truck, the brake shoes being used as current collectors.

Axleboxes or journals are among the most difficult parts to make and are best described by the use of drawings. Care in marking out is essential, and as at least four per car are required the easiest method is to solder four thicknesses of tinplate or brass together and fretsaw them to the correct profile. The journal itself is made up of three layers of brass and the whole lot sweated together after bending to shape.

Allowance for the radiused ends can be calculated as follows: Spring cups are 0·140in diameter for 3/8in scale, and as only half a circle is required the correct length is simply 1·571 x 0·140 or 0·22in for each cup. The alternative is a much more complicated method involving the use of metal moulds and white metal castings which ensures that each journal is exactly like all the others and requires only the axle hole drilling.

The diagrams, Fig 45, show the type of Brill 21E wide wing

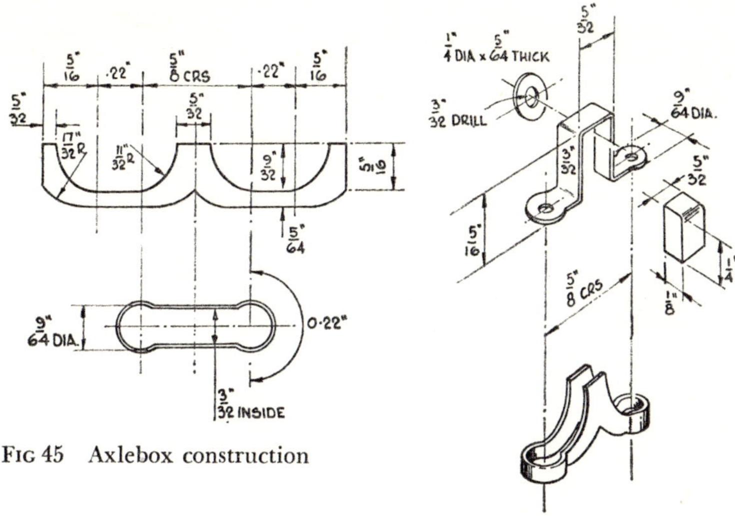

FIG 45 Axlebox construction

axlebox. If the 21E type is required the only dimension to alter is the 5/8in crs to be reduced to 13/32in below the side frame; all other dimensions remain as drawn.

The truck sides are much easier to produce either from tinplate or brass by sweating four layers of metal together and then marking out and sawing to shape. It is possible to use only two strips of metal without loss of appearance and to solder the spring pins to the back of the side frame at 3/8in centres. If realism is to be achieved do not forget that almost all older truck side frames had the leaf spring cantilever sloping slightly upward to the extent of 1in on the prototype or 1/32in on the model; again these ends have a flat plate at the extremity on which to mount the leaf springs and this can be either left on the side frame in the shape of a lug to bend flat or soldered on afterwards.

Dependent upon the gauge chosen will be the length of any cross bracers or transomes (see Fig 46). For 3ft 6in gauge (nominal) these will always be made to produce a centre-to-centre distance of the truck side frames of 5ft 2in full size ($1\frac{15}{16}$in) and for 4ft $8\frac{1}{2}$in gauge this is $2\frac{1}{8}$in or 5ft $8\frac{1}{8}$in on the prototype. These are almost always made from flat strip of 3/32in x 1/32in cross section with each end bent through a right angle for fixing to the truck sides by two bolts (16BA). Coil springs are not difficult to obtain but may need cutting to correct length on assembly. These

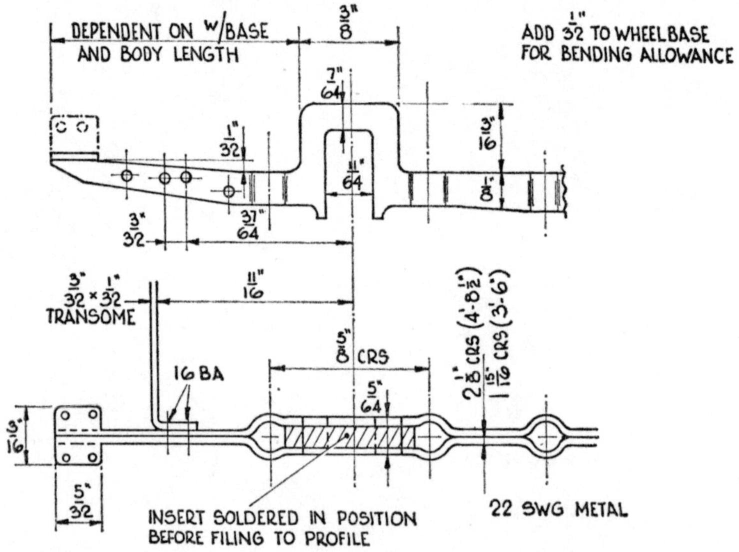

FIG 46 Truck side

ought to be 1/8in diameter and 9/32in long, unless the car body is of heavy construction when they may need to be increased to 5/16in. This will allow the body to sit at the correct height on the truck. Care should be taken in selecting coil springs since springs which are too strong will defeat any attempt at getting them to work, and if they are too soft they will close up until the car body rests on the top of the truck side. It is considered better to use a spring which is too light or soft than one which is too strong. A second and smaller diameter spring can be added inside the other if required. The reason for this is that, in addition to the coil springs, there are usually four leaf springs to share the load, as in the 'Peckham' cantilever standard tramcar truck shown in Fig 47. If the leaf springs are to be dummy or non-working then allowance can be made for the car body to be carried on the coil springs so that there would normally be a gap, say 1/32in, between the spring pads on the truck top-plate. The leaf springs are attached to the cantilevers by inverted 'U' bolts made from 24 or 26 gauge wire. The gap so left is completely hidden by the spring rubbing pads which have little wing-like ear flaps on each side.

The truck top-plate is the flat steel bar along the top of each

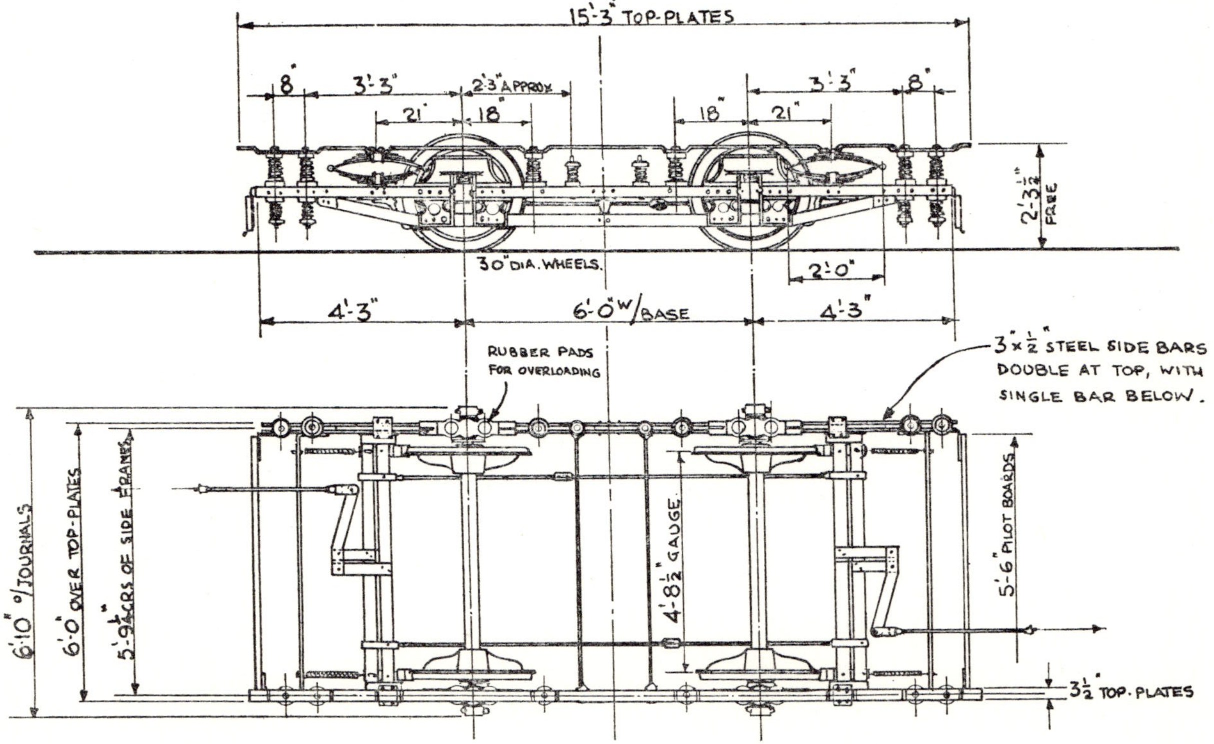

FIG 47 Peckham cantilever tramcar truck

side and which is attached to the car body, thus uniting truck and car. With a fully sprung truck it is almost impossible to get this thin flat strip to remain flat if there is the slightest unevenness in the spring pressures, so to achieve this it is necessary to add a stiffener to each top-plate in the form of a vertical rib soldered on the top face, and which also serves as a location face for the spring pins which pass down the centre of the coil springs and are soldered to this inverted 'T' section, Fig 48.

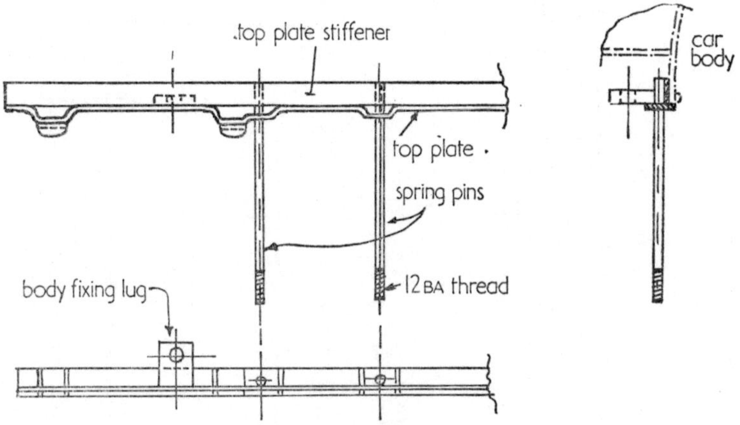

FIG 48 Truck top plate

It is advisable when assembling the truck to do it in the upside-down position, using the two vertical ribs as the base, standing on a piece of plate glass in the absence of a surface plate. Cross bracers can be used across the top-plate, but if these are used care should be taken to place them clear of the platform beams, which project beneath the car body, and also, if the motor projects through the floor, care should be taken to leave clearance for this. Temporary cross braces may be tacked on at the assembly stage to keep the truck sides vertical and then removed later or used as the brackets for bolting the truck to the body. Great care is needed in assembling fully sprung trucks if the wheels are not to 'rock' when placed on the track and 12 BA screwed rod is ideal for spring pins since the springing can be adjusted to level by tightening or slackening the nuts beneath the axleboxes, but there is nothing to prevent a blob of solder being used instead of nuts.

In later years, due to the heavier weight of top-covered cars, very many trucks had truss bars passing from one end of the truck to the other. These consisted of special brackets at each end with diagonally sloping tubes down to the outer extremities of the axle-boxes. The horizontal bar beneath each axlebox was flattened to pass the spring pins and was a round bar between the innermost axlebox springs.

Almost all trucks had a 'pilot' board across each end about 4 to 6in above rail level and these were supported on slender curved arms attached to the cantilevered ends of the truck. These are usually better made from slightly heavier gauge metal as it is the pilot board which gets the knock when a model derails and these brackets need to be more robust than the other parts for this reason. They are difficult to produce since there is a lot of tedious work involved in drilling holes then sawing to outline and filing to shape. The addition of half-round brass wire to give the correct 'T' section is made after soldering to the truck side frame cantilevers. Some manufacturers used a much simpler bracket; examples are Hurst Nelson and Boving Engineering, both of whom used flat plates bolted or riveted to the inside of the side frames.

Whether a truck is fully sprung or not the foregoing is applicable, the only difference being that for an unsprung truck each part can be soldered as the truck is assembled or the parts above the truck side can be 'solid' leaving only the axleboxes sprung. This is also a very satisfactory method and has been proven over many years.

Bogies are usually less complicated to make than four-wheel trucks and a great deal of the more complex construction is well hidden from view and can often be omitted without loss of appearance. A very important and often omitted part is some means of preventing the car body rolling sideways through relying only on a centre pivot pin. The addition of radius plates at each side of the centre pin is quite easy and will ensure that the body remains upright under all circumstances.

The use of springs over the journals is optional and any form of springing can be dispensed with entirely if the bogie side plates are made to flex or pivot about a central point.

Most types of bogie are easier to construct as has been said, but they have the disadvantage of requiring two per car therefore double the work, but the biggest problem is the motor drive which usually means one bogie pivot bolster has to be cranked upwards to give a clearance for the motor. Another disadvantage is that the car has a drive on one axle only which can all too easily cause excessive wheel spin and loss of motion on the sharper curves.

The alternative method is to mount the motor on the car underframe, using a motor with a double shaft extension coupled by flexible drive shafts or Bowden cable to a worm drive unit on one axle of each truck. If it is possible to incorporate a flywheel on one end of the armature shaft, operation will be much better. On four-wheel trucks the mounting of the motor is dependent upon the type and size of motor it is intended to use and again it is desirable for the best performance to incorporate a flywheel on the motor armature shaft and utilise a worm drive to one axle only (Fig 49). This arrangement is quite adequate for most model tram-

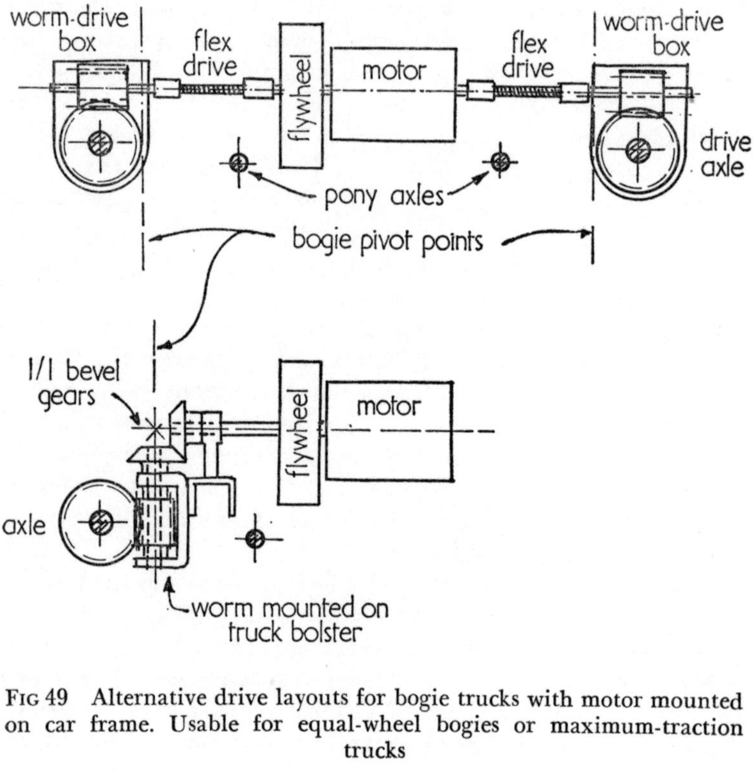

FIG 49 Alternative drive layouts for bogie trucks with motor mounted on car frame. Usable for equal-wheel bogies or maximum-traction trucks

cars up to $3\frac{1}{2}$lb weight being operated on gradients up to 1 in 10. Gear ratios are important in making a model tramcar and nothing is worse than a model tramcar doing a scale 100mph. Remember that the tramcar was once considered to be the fastest vehicle on the road usually travelling at speeds of from 8 to 15mph. Experience has shown that a model car with $1\frac{1}{32}$in diameter wheels (scale 33in diameter) needs a worm and wheel ratio of 40/1 using a 24 volt, 5 or 7 pole armature motor. If a modern type of car is to be made then the speed should be a little faster. The wheels, however, are likely to be $\frac{27}{32}$in diameter (scale 27in diameter) and a 25/1 gear ratio is nearer the scale speed required. Both the above ratios are available commercially at reasonable prices.

The choice of motor is one that can be decided by the builder. If commercial model railway equipment is to be used as the source of power then the choice will almost certainly be 12 volts dc; if on the other hand 24 volts is used there is still a fair choice of ex-government equipment on the market. The higher voltage has the advantage of requiring less time spent in cleaning track and overhead lines, and gives far greater tractive effort.

It is advisable to fit suppressors against interference with television. It is, however, likely that there will be a great amount of sparking caused by the small area of contact between trolley wire and trolley head, where it is unfortunately impossible to fit any kind of suppressors which can be hidden.

Where scale speeds are aimed at using a worm drive then a flywheel on the motor is an absolute must if the car is to coast to a stop instead of virtually standing on its end each time the power controller is turned off. It is advisable when buying a motor from a model railway supplier to purchase it with a flywheel already fitted so that it can be dynamically balanced in position. A flywheel which is only slightly out of balance can cause a most unpleasant noise when driving a model. The improved performance of a fitted flywheel is well worth the extra cost or time spent adding one.

The worm drive mounting on the axle will have to be made to suit the type of motor purchased but one is obtainable with a formed brass 'U' shaped bracket which slides onto the drive axle and is located by the worm wheel which is pressed onto the axle. A similar form of fixing is equally suitable for use with larger

24 volt motors, or a more elaborate enclosed worm box can be made.

Yet another method available for bogie cars is the use of an intermediate bevel gear using a rigid motor extension shaft on which is mounted one of the bevel gears. The other right angle bevel is mounted on top of a vertical spindle which also forms the pivot pin of the bogie. Immediately below this is a worm drive with the wheel mounted on the drive axle. The only deviation from scale mounting is due to the necessity to alter the scale distance from the drive axle to the centre of the pivot point of the bogie, which is determined by the working centres of the worm and wheel. The layout is shown diagramatically in Fig 49.

This idea is usable only on certain types of maximum traction bogies and could not be adopted where the pivot point of the bogie is directly above the drive axle as in the Maley & Taunton trucks used on Birmingham Corporation bogie tramcars and also referred to as 'Burnley' bogies, after the system on which they were first used. It will be seen that in every case described the motor is so mounted on the truck (or car frame in the case of bogies), that the scale location of the lower deck floor of the car body must be sacrificed, but allowing for working clearance it is possible to keep the 'floor' low enough for the seats to be mounted directly on to the floor without appreciable loss in appearance.

Body Construction

In the smaller scales the type of material used and the method of attaining rigidity and appearance do not differ greatly from those used in the larger scales. It is intended in this section, however, to describe a typical 3/8in scale construction using tinplate, solder, and glass for windows.

Experience has shown that the best method of constructing the lower deck saloon is to produce one side and one end or bulkhead as one unit complete with all seats and mouldings so that when final assembly is made the two 'L' shaped assemblies fit together to form an open-topped box without a bottom. No fixing screws or bolts are used other than fixing lugs pre-soldered to the inside of the rocker panels ready to accept the screws which hold the floor, which is inserted from the underside. The top of the saloon is made complete with cornice brackets onto which drops the saloon ceiling, again a drop-in fit without fixing screws. The in-

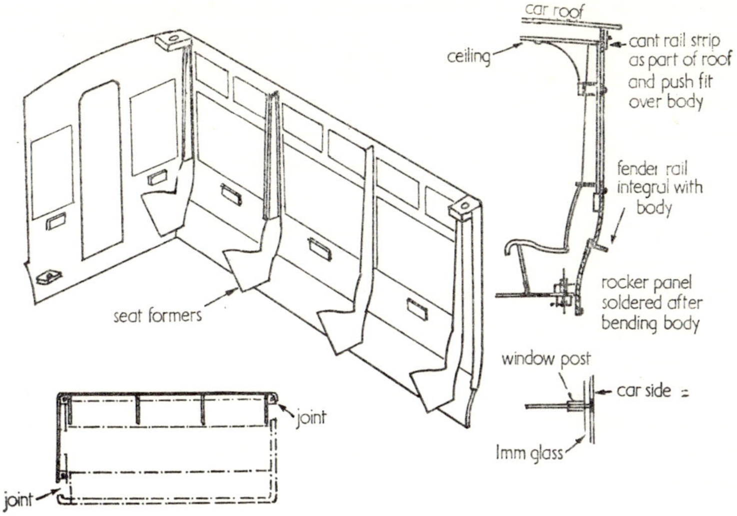

FIG 50 Body construction

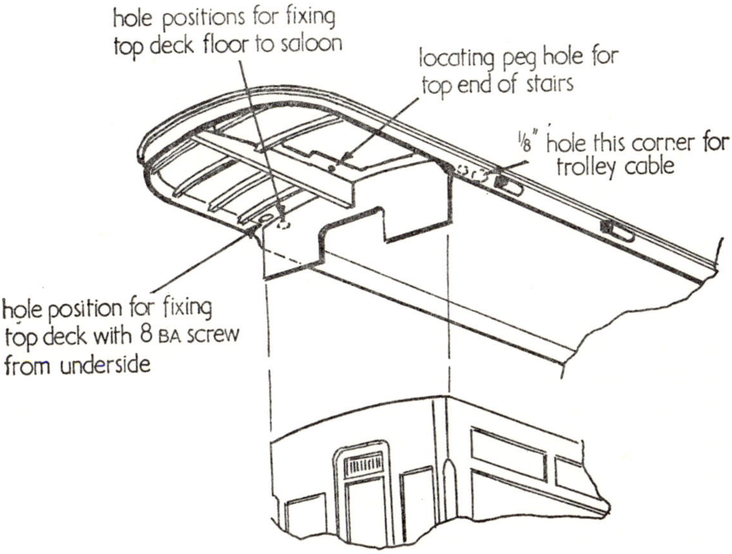

FIG 51 Method of assembling body

terior painting is done before assembly and this is one of the reasons for this method of construction. The location of the top deck floor is ensured by making use of the cant rail depth and bulkhead facing boards which form a shallow sided box into which fit the top of the saloon sides and ends See Figs 50 and 51.

The top deck floor is fastened by screws to four fixing lugs in each corner so that the lower saloon, floor, and top deck floor are now as one unit and ready for painting. If tinplate is used frequent washing in soapy water is essential after soldering if the appearance of rust is to be avoided. One or two coats of primer may be applied at this time if desired.

Exterior mouldings should be added wherever possible before assembly and as the vertical joint at two diagonally opposite corners needs to be covered, there is no better way than using the corner cappings to half-lap the joint in the body, exactly as in the prototype car.

Screwed assembly is more time consuming but is desirable if glass is used for windows, for if a window should become broken at any time the model can be dismantled and a replacement fitted without ruining the model.

Although metal body construction has been dealt with, other media are equally adaptable to the same technique of construction and at assembly the joint may be glued or cemented together and clear Perspex can be used for windows in place of glass. A combination construction is ideal where opening windscreens or drop lights are made to work. The window post and seat former at the corner nearest to the body joint serves a twofold purpose: (1) to provide a rigid support for the seat end and the end window, and (2) being about 1/8in from the end of the body, it provides a cable channel through which the wire from the trolley passes. The car platforms are attached by screws to the underside of the car floor using blocks of metal to pack the platform beams to the correct setting.

Construction of the platforms is made as near as possible to the prototype. There are four beams bolted to the body headstock beams by suspension bolts having plates beneath. The inner ends of the platform beams usually go under the body about 3ft 0in on four-wheel cars and the ends rest under another cross member just at the outer extremities of the wheels. Construction for bogie cars differs mainly in that the two outermost platform beams are almost always cranked outwards from the headstock to fix to the underside of the body sills whilst the two innermost beams stop short to clear the motors on maximum traction bogies.

In model form the platform beams can be made from flat brass section 7/32in deep x 1/16in thick set on edge and soldered to the underside of the crown plate and platform floor.

If brass channel cannot be obtained to the correct size the fenders can be made up by sweating four thicknesses of metal together and fretsawing them to shape so that when separated the top and bottom flanges are made for the channel section. The vertical part or web is bent to the shape of the flanges and soldered in position on the ends of the platform beams. The fender could also be bent from a piece of brass or steel flat metal.

The platform floor, that is the area between the headstock and crown plate, is always made of thinner material, approximately 1/32in thick so that there is a difference in depth between platform and crown plate and this is put right when the wearing strips are added to the top face. The dash plate or vestibule is soldered or cemented to the edges of the platform floor.

When fixing the platforms to the car body a very slight upward slope towards the fenders is necessary. The platforms are easily fixed to the car body floor by two screws into corresponding blocks soldered beneath the floor so that they can be quickly detached if required.

Working lifeguards are a very nice refinement but can be more trouble than they are worth on a small working model and for this reason they are best made as fixtures permanently soldered to the underside of the platforms.

Handrails can be made from 18 swg brass wire soldered in position on the stairs, and where joints occur a small hole for location is drilled in the body, or if attachment is to a stanchion post the adjoining wire should be pushed halfway through the holes so that no joint is visible when assembly is complete. It is worth contacting a model boat supplier to inquire if suitable stanchions can be purchased before starting to make one's own.

The top deck construction on covered-top cars follows the same pattern as the lower deck, except for fixing the roof to the sides, where any screws would completely upset the finished appearance. Therefore blocks of metal are soldered to the underside of the roof and matching lugs are sweated to the top corners of the sides so that all screws are inserted from the underside and so are invisible from the outside.

Nothing has so far been said about fixing the top and bottom decks together. This is indeed not an easy job to accomplish and there are several ways of doing it.

1 To pre-assemble and complete all painting except at four points so that the complete top deck can be soldered in position. This is quite satisfactory but once soldered it is a messy job if the two halves ever have to be separated.

2 To solder any desired number of horizontal lugs to the bottom of the sides, and all pointing in the same direction, and an equal number of loops to the top deck floor so that the lugs are a good push fit into the loops. All that is necessary to separate the top and bottom decks is to push them in opposite directions until the lugs and loops disengage. Accurate positioning is essential.

3 To add four lugs or small blocks of metal to the outside corners of the top deck bulkheads so that screws can be inserted under the platform canopies where they are usually fairly well hidden (See Fig 51).

It will be seen from the methods described that final assembly is a little like building a house from the chimney pots downwards in order that all fixing screws are out of sight, yet the model can be dismantled and reassembled at will.

Top deck tip-over seats are fiddling items and design varies from contractor to contractor but the most tedious to produce are the type known as the 'dry flap' seat where a narrow portion of the seat could be flipped over to form a dry seat after a shower of rain. If desired the seat and legs can be formed from one strip of metal, not forgetting that due to the camber on the top deck floor the inner leg will be shorter than the outer leg. For top-covered cars the leg nearest to the car side was omitted and the seat end rested on a batten of wood screwed to the inside panels of the top deck. To add the realism of a wooden seat drill two, 1/8in holes in the metal seat bottom and cement the shaped wooden seat top. To make sure that the two never come apart cement beneath the metal seat bottom a strip of thin card long enough to cover the two small holes, thus forming a sandwich construction.

To ensure uniformity of all seats it helps if an old hacksaw blade can be filed to the profile of the wooden seat bottom and then scraped along the previously prepared strips of wood. This operation is easiest carried out before cutting off to the required seat lengths. Tip-over seat backs are stronger if made up using a narrow strip of metal with two pieces of wire soldered to it to form the supports. The bottom 1/16in of each wire is bent inwards and simply sprung into the pivot holes near the bottom of the seat legs. Once assembled each seat should be soldered or bonded to the car floor and painted in situ if on open-top cars, or soldered to the top deck sides on top-covered cars.

For open-top cars the best method of construction of the decency boards and hand rails is to assemble as one complete unit with four lugs arranged to coincide with the holes used for screwing the top floor to the saloon. The 8 BA screws are hardly visible on the completed model.

The spiral stairs are often thought of as being the most difficult parts to make but care in cutting out the treads and risers is the part of the work which will make or mar the effort.

When assembling the stairs, a wooden former is advisable so that the angle is consistent when soldering the metal parts together. This former can be made from a piece of wooden dowelling 1½in diameter cut to the length required, equivalent to the height from platform to top deck floor, and then sawn in half longitudinally where 180 degree turn stairs are called for. The length is then marked out in equal divisions corresponding to the number of risers required: sometimes eight but more often nine steps (ie seven treads plus top deck floor or eight treads plus top deck floor). Next divide the half-round end into seven or eight angles and mark these longitudinally along the former, then proceed to saw-cut the steps. Once this is made the metal tapered step treads are marked out and a parallel piece left on one end which is bent at right angles to form the vertical riser up the back of each tread. Lay the finished angle on the first step of the former with the second step above it and solder the two together and so on until the correct height and spiral are attained.

Next cut a strip of thin metal 5/16in wide and bend it along the stepped outside curve of the former leaving about 1/4in overlap at both ends. When this is curved correctly, solder it to the steps to make the curved outer stringer. The stairs will now be quite rigid and can be handled without fear of distortion. Finally add the inner stringer by twisting a narrower strip of metal 3/16in wide between the fingers, then soldering this to the inside of the steps. All car stairs are finished off by the addition of half-round beading along each edge of both inner and outer stringers. Leave cutting the stringers to correct length until actually fitting to the model. At the back of the bottom riser solder two short bits of wire to locate the stairs into two small holes in the platform floor. A similar pair of small wire pegs at the top end of the stairs serves as a boltless fixing for the stairs under the canopy.

Wiring and Power on the Model

In the smaller scales lighting a model is the exception but there is nothing to prevent lighting being added if sufficient power is available to feed a series of bulbs in addition to the motor. The difficulty is more likely to be the physical dimensions of the bulbs rather than adequacy of the supply, indeed several bulbs in series could well be used as a resistance for 12 volt motors operating from a 24 volt supply. There are many arguments for and against adding lights to small-scale models. The method of lighting is basically the same as used for the larger scales. Power supply on the model will depend on whether or not the overhead line is to be energised. Cars can be arranged for two-rail current collection with track sectioning and have a trolley purely as a dummy unit. An alternative is to use the trolley as a means of superimposing a low voltage alternating current for lighting purposes only. The other method is to use a system such as used by 'Trix' trains where the wheels are insulated and the third rail (the trolley-wire in this case) provides the means of operating two cars at different speeds on the same track.

Track Construction

The appearance of a model tramway track is very vital to any layout and is often the deciding criterion of just how good or bad a layout is. There is a tremendous amount of work involved in laying good track, but on the other hand, it is equally difficult to lay bad track when it is desired to copy some of the poorly laid and maintained prototypes found in many parts of the world.

A well-tried and comparatively easy way to produce rail is to use 0 gauge fine-scale flat-bottom rail with the 00 gauge equivalent soldered outside into the web of the larger section so that the groove of the tramway rail is formed by the flat base of the 00 rail soldered in a horizontal position. Lay the track formation and either screw or spike it to the baseboard, then add the check rail in situ; a blob of solder about every 3in is all that is required. The only drawback to this method is that it produces a slightly overscale groove width but only the fanatic would notice a discrepancy.

The alternative is to lay the 0 gauge fine-scale flat-bottom rail and to then solder in a small-scale section of angle which can be obtained if you know a sympathetic tinsmith who will guillotine it to width and roll it in the shape of a letter 'L'; this then needs soldering into the web of the running rail.

Laying the track is an easy task if small 3/8in x 1 countersunk head woodscrews are used so that the countersunk parts of the head can be used for making 'micrometer' adjustments to the gauge of the track as required. It is important to use a well-made track and wheel gauge, dimensions of which are given in Fig 52

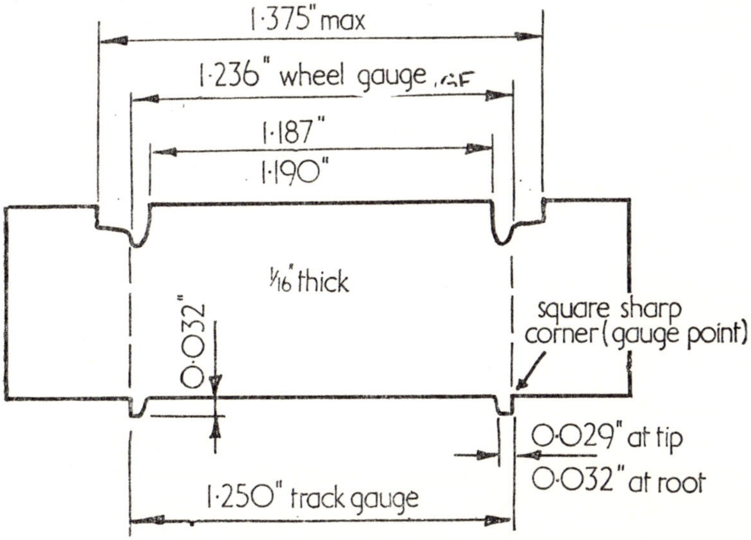

FIG 52 Track and wheel gauge. Dimensions shown are for gauge 0

for 0 gauge. This same gauge is also a means of ensuring that the 'groove' in the track is of the correct width. It is often necessary on sharp curves to increase the width of the track groove, exactly as was done on full-size track.

Full-size track has flat steel tie bars every 5ft 0in but unless exposed track is to be left these can be omitted, since they are invariably buried in the paving or ballast and therefore do not show.

Points and crossings take more time than their railway counterparts as the grooves are continuous. Points can be built up using three layers of metal, either mild steel or nickel silver, a thin layer being used for the base which is slightly wider than the rest, and provides the means of fixing to the baseboard. On top of that is the middle layer of 16 swg and on top of that are the 18 swg strips to form the running surface of the rails. To the middle inner edge of the point body is soldered a 7/32in wide strip of metal to form the check rail, which is 'kinked' to accept the toe of the point blade. The running surface is also 'kinked' in the reverse way for the same reason. Many tramway systems used points which had one moving blade only, the other being fixed with no moving parts whatever. Where double moving blades are used care should be taken to see that one is left-handed and one is right-handed and the two are joined together with a cast trough into which fits the tie rod and operating mechanism if automatically operated from the car. This consists of a rotary solenoid mounted in a pit between the points and connected to the point blade tie rod. Where spring points are used, such as on single line with passing loops, the point blade(s) can be made to operate by extending the point blade pivot (or heel) down through the baseboard and adding a light spring or a rubber band to keep the blade over to one side so that all approaching cars take the same direction.

Where rails cross there is usually a cast 'frog' which is easily made from a piece of 3/16in thick steel or nickel silver. To make the groove, simply use a hacksaw with a coarse blade and saw to a depth of 1/16in at the ends, rocking the saw to provide a groove only 1/32in deep at the point where the grooves cross each other. This has the effect of allowing the car wheel to run on the flange so, in theory, eliminating the bump which would occur if the tread had to jump the crossing groove. A thin piece of metal soldered beneath will bring the frog to the same height as the rail and the projecting edges can be drilled for screwing to the baseboard. Each frog is best insulated from each adjoining rail by inserting a thin piece of plastics material across the rail ends. This prevents trouble when wiring the track later.

Paving of the street tramway track can be put into three categories (1) sett paved (2) tarmacadam (3) reserved sleeper track used on tracks segregated from the highway where ballast is generally used to a depth of one to two inches below the running surface. Occasionally the ballast was laid to conventional railway practice, eg, Blackpool, Fleetwood, Grimsby and Immingham, and Huddersfield, but these are exceptions to the general practice.

If sett paved track is used Polyfilla is as good as anything as this can have suitable pigment added to obtain the correct reddish-brown granite colour, although grey granite blocks have also been widely used. When the filling has almost dried, there comes the tedious job of scribing out the setts. First job is to scribe lines at right angles to the rails, and then go along each space, marking the setts in a staggered form. Each sett will vary slightly and spacing for the rows across the track should be at 1/8in for 3/8in scale or 4in full size and each sett will be 7/32in to 9/32in long or 7in to 9in full size. So that for 3ft 6in track gauge there will be 5½ setts between the check rails and for 4ft 8½ gauge there will be at least 6½ to 7 setts. The obligatory 18in of paving outside the rails would consist of 2 or 2½ setts dependent upon size irrespective of track gauge. At points and crossings several arrangements of sett paving were in use but one will never be far away if the paving is laid at right angles to the straight track of the turn out and the paving for the curved line will be radial to the curve once outside the 4ft 8½in straight line.

A much less used system of paving is to lay one line of setts in line with and adjacent to the rails and then fill in with tarmacadam. An essential detail not to be overlooked whatever style of paving used, is the position of drain boxes for taking away storm-water where track is in a hollow or on any gradient. These are usually opposite each other in each rail and constitute a square iron box bolted to the back of the check rails. Water in the groove runs out to drain through a slot about 6in long in the bottom of the groove.

The standard length of prototype tramway rail is 60ft so if realism of sound is desired a file cut should be made in the rail every 22½in for 3/8in scale. The use of outdoor track in the smaller scales is very much an unknown quantity but of the few known to exist none have used scale track, and none have scale trolley-gear, but they do work and may well be worth further experiment.

Nothing has been said of track gauges used and a few lines on the subject may not be out of place. In Britain the three gauges most commonly used were 3ft 6in, 4ft 0in, and 4ft 8$\frac{1}{2}$in although metre gauge is used on the Manx Electric and variants like 4ft 7$\frac{3}{4}$in in places like Glasgow, Hudderfield, etc, where railway wagons would traverse tramway track. In 3/8in scale 1$\frac{5}{16}$in is equal to 3ft 6in gauge so it is suggested that the nearest commercial standard is 1$\frac{1}{4}$in gauge which is a scale 3ft 4in which is suitable also for metre gauge. It is fully appreciated what the 'dead scale' enthusiasts will have to say about this; nevertheless the choice allows commercial equipment to be purchased which would otherwise be unavailable except to special order. 4ft 0in gauge was seldom met with outside North-east Lancashire and three systems in the West Riding, and is perhaps best converted to 1$\frac{1}{4}$in gauge. There are a number of models of 4ft 0in gauge cars running on 1$\frac{1}{4}$in track which do not look out of character in any way. Standard-gauge track in 3/8in scale measures 1$\frac{3}{4}$in and is therefore true gauge.

One of the objects of this book is to offer a work of reference for tramway modelling and it should therefore encourage the use of standard dimensions for track and wheels.

The construction of suitable baseboards offers considerable freedom of choice, provided rigidity is observed. A suitable frame section is well seasoned 2in x 1in softwood for the base frame covered with 1/2in tongued and grooved boards. Chipboards provide an alternative to boards, but where rails are to be screwed direct to the chipboard screws which are too short may tend to break out if the baseboards are handled too often. The overall size of any baseboard will be governed in all probability by the size of access to the house loft or size of the layout room. Unlike model railway baseboard there is an additional complication to contend with on a tramway model baseboard and that is the allowance for height of overhead wires and poles, and the provision in the baseboards themselves of suitable blocks of wood beneath the boards to provide rigid fixings for the trolley standards. These need to be added at the construction stage of the baseboard.

Sections of baseboard need to be accurately located, one with the other, and 1/4in carriage bolts are quite successful for this purpose if made a light push fit. Provision for section wiring is also required and the common plastics cable connector blocks are ideal and cheap. Multiple-pin plugs and sockets are perhaps better if a large number of ends are to be used, such as a remotely situated depot layout. If the tramway tracks are to be paved then rigidity of the baseboard is vital if the plaster filling is not to crack, especially where a layout is to be portable.

Overhead Wire Construction

This is a most fascinating aspect of tramway modelling and is well worth the effort involved in obtaining satisfactory operation irrespective of whether the line is to be a power conductor or just for appearance. Copper wire is quite satisfactory but in 3/8in scale 26 swg phosphor-bronze wire is the best as it will not stretch when tensioned, as copper will, and it is far easier to 'string' since it does not kink easily. Straight-line hangers are used where the line is straight and bridged type where the wire goes round a curve, usually where the angle of the wire changes more than five degrees. This is not vital but does add considerably to the realism. Curved line hangers have the disadvantage of being very difficult to produce whereas straight hangers can be obtained either as small white porcelain rings or model yacht fittings made in wood may be used. Each has a hole through the centre through which a 12 BA cheesehead brass screw is inserted and screwed in position and a groove around the edge provides a mounting for the span wire. Into the screwdriver slot of the 12 BA screw should be soldered the ear which carries the trolley-wire.

At junctions where the wire divides, frog pans are made from 0·008in thick copper or brass strip with the trolley wire carefully soldered in position. For assembly this must be carefully marked out on a piece of paper on the workbench and sufficient trolley wire cut off to reach to the nearest insulator in each direction. The frog pan metal is thin enough to be bent into the shape of a trough, the sides of which guide the trolley wheel when in position.

To form the junction in the wire, solder the wire which enters the frog and leaves at an angle in position first. Next solder the continuation of the straight run through leaving 0·015in gap at the point where the angle occurs in the first wire. To ensure that the trolley wheel flanges will not jump over the angled wire file a groove to allow the wheel to run straight through (Fig 53).

The position of the frog to 'pull off' at the correct point over the track is easily determined by the position at which the car trolley begins to leave the straight line: once the trolley begins

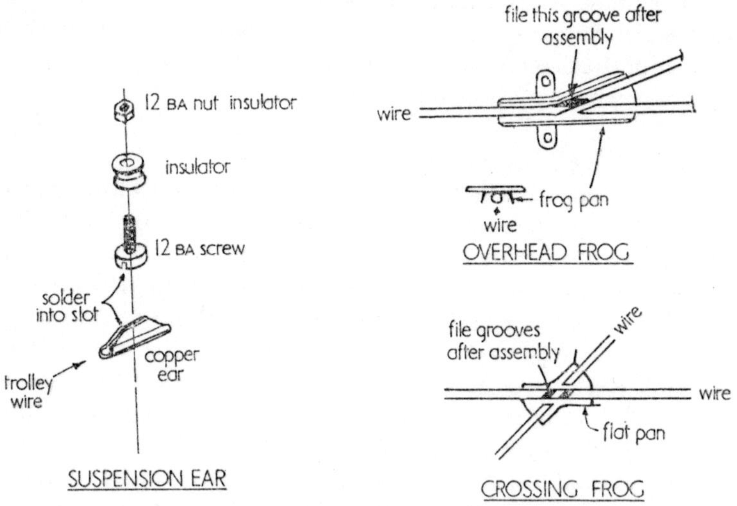

FIG 53 Overhead line details

to be dragged sideways it will take the first opening it comes to and thus pull off at the desired point without the use of any moving parts. Where a trolley reverser is to be used the same construction is applicable except that the filed gap in the angle wire is filed only half the full depth and this will be sufficient to guide the trolley wheel through the angled pull-off when propelled. If this procedure is closely followed only a light upward pressure is necessary on the trolley-head. Too much pressure will stop the wheel entering the trough and in all probability will result in dewirement and damage to span wires. Spacing of prototype traction poles varies from 75ft (28in) to 120ft (45in) dependent on circumstances so that if poles vary in spacing it is only what happened on the prototype. Side poles and bracket arms were widely used and the bracket arms went in increments of 2ft 0in in length from as short as 4ft ($1\frac{1}{2}$in) to 22ft ($8\frac{1}{4}$in) which was the maximum permitted length. Ornamental scrollwork is very picturesque though not essential. Poles themselves were built up, usually by using four sections of tube, each smaller than the base section and pressed firmly one inside the next. More often than not ornamental collars adorned each joint and an elaborate cast-iron skirt was fitted at pavement level. Traction poles always had a backwards rake on erection to counteract the weight of the overhead.

Trolleys

The trolley is a most important item on any model since it not only provides the means of current collection, but is one of the first things to be seen since it is on top of the model. Many otherwise excellent models have been spoiled because the trolley base is held on by a big square nut with a screw end projecting up still further. It is therefore worth the extra effort involved in arranging the fixing with a special screw cap or if no lathe is available a countersunk head screw.

The trolley will of course need to be adequately insulated if used for current collection and this can be done by using a wooden or plastics trolley-plank screwed to the model roof by 16 BA nuts and bolts. The actual swivel base is then in metal and attached to the trolley-plank. The cable to the motor is passed through a small hole in the car roof and trolley plank and soldered to the swivel base. The cable should have enough length to be laid between the car roof and the interior ceiling to the corner post of the body and down to the motor (Fig 54).

The trolley standards for open-top cars were produced by a number of manufacturers but those which had internal tension springs are easy and simple to make. The base section about 7/8in long and 11/32in wide shapes off to 3/16in round at 9/16in above the floor and the mast is 3/16in diameter for $1\frac{3}{8}$in. The base can be made from Bakelite or similar material with a 5BA tapped hole in the circular top into which is screwed the mast. The top of the mast should be drilled 1/8in diameter to a depth of $1\frac{1}{8}$in and this will hide the springs attached to the trolley boom. The trolley and its hinged base simply push into the hole and are free to turn through 360 degrees. The mast head is assembled with a loop of soft wire attached to the hinge and tensioned by soldering the bottom ring under the spring as desired.

The trolley boom is made from 3/32in diameter steel rod or bicycle spoke turned to a slight taper towards the head. The swivel head of the trolley is a delicate item to make, especially the forked end in which the trolley wheel fits as this is 'waisted' near the swivel bearing, leaving very little metal at this point. Extra care is therefore necessary at this stage. A lathe is absolutely essential at this stage and if a number of trolleys are required then form tools are strongly recommended. These can be made from silver steel and afterwards hardened for use in the lathe.

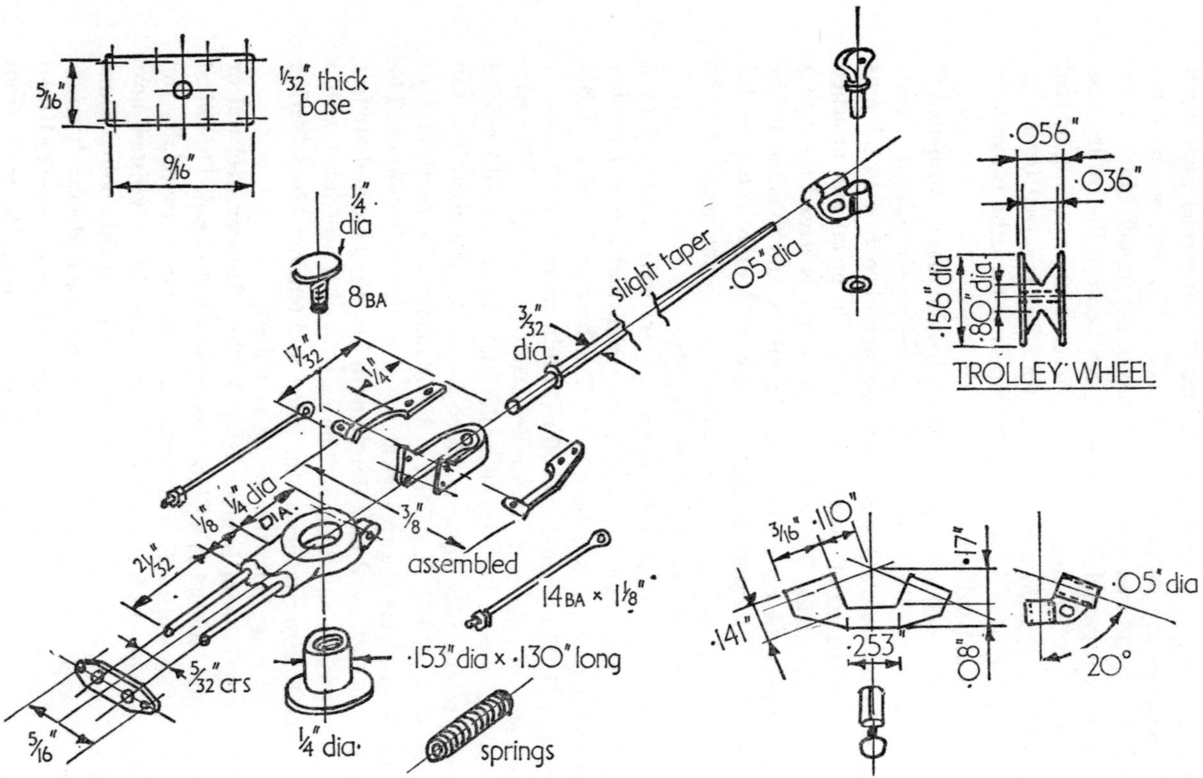

Fig 54 Details of trolley for top covered car

To produce the head swivel bearing 0·010in thick brass strip is used and formed by simple jigs held in a pair of pliers.

Painting

The same methods of painting are applicable to the smaller scales as to the larger ones but more care and attention are required to ensure that the paint is used sparingly and not allowed to thicken at the corners, as delicate mouldings are easily lost if paint is applied too thickly. The extensive variety of commercially produced lining and lettering available today makes the task of titles, numerals, and notices far easier than was the case some years ago.

It is necessary to emphasise again that the success of operation and running a model tramway system relies on good electrical conductivity and nowhere is this more apparent than in the smaller scales. The two cardinal principles for success are the same as for the larger scales, the first being the weight of the model. A lightweight model will cause endless trouble by its very lightness which does not ensure good electrical conductivity on the rails, especially if there happens to be any dust settlement on them. Fortunately the weight of the motor, gearing, trucks and ancilliary equipment in the smaller gauges also ensures good electrical conductivity. The second essential is adequately powerful springing in the current collection whether this is of the trolley-pole bow-collector, or pantograph type.

The information given in this chapter is derived from methods which have successfully been used over many years by a considerable number of modellers in this scale. Naturally there is scope for the use of many other ideas and the individual modeller will doubtless have the necessary enterprise and ingenuity to apply some of his own.

16 TRAMWAY SOCIETIES AND ORGANISATIONS

The national societies and organisations connected with electric tramway interests are widely known and membership is usually open to all enthusiasts. The seeker after knowledge, good companionship, and activities in this field can rely on obtaining full value for his modest subscription especially if he is willing to give as well as to take.

Most predominant for modellers is the Tramway & Light Railway Society founded in 1938 which has as its objects (i) the bringing together of those interested, professionally or otherwise, in light railways; (ii) the encouragement of research into and study of all facets of tramways and light railways; (iii) the collection and collation of tramway and light railway archives and other material, the maintenance of a library and the publication of literature, photographs, and drawings of tramway and light railway interest; (iv) the encouragement of tramway and light railway modelling and the promotion of and participation in exhibitions; and (v) the holding of meetings to consider matters of tramway and light railway interest and the arranging of visits to tramway and light railway systems and places connected therewith.

The Society has authoritative records of tramways, past and present (including overseas tramways still operative) in its archives. Most items are available to members upon application to the Librarian and other items not for loan may be examined by prior appointment. A wide selection of tramway photographs is always available at reasonable prices at most general meetings and exhibitions where the Society is exhibiting or directly from the Photographic Secretary who can also supply a copy of the photographic catalogue.

An illustrated magazine *The Bulletin* is issued free quarterly to all members. It contains articles on tramway systems, model engineering, reviews of general meetings and new publications, Society news and announcements, and a vigorous correspondence column. A newsheet *Tramfare* is also issued to members.

Keen model-making groups exist within the Society membership working respectively to scales of 3/4in to the ft, 3/8in to the ft, 0 gauge and 00 gauge. The Society owns 3/4in to the ft scale layouts in standard (4ft 8½in) and narrow (3ft 6in) gauges which are regularly demonstrated at exhibitions, providing members with opportunities to exhibit and run their own models. Technical advice and information on bodywork, mechanical, and electrical aspects can be obtained, from recognised experts, through the Model Engineering Secretary.

Informative booklets on notable tramway systems have been published and are available from the Publications Officer. Regular monthly meetings are held in London when illustrated talks are given on every aspect of trams and tramways. Other interesting activities include film and slide shows, general knowledge and photographic competitions. A group of the Society exists in Birmingham where a flourishing model engineering section has its own layout in narrow gauge for exhibition purposes and general use. Meetings are held regularly in Birmingham. Close liaison is maintained with other tramway societies to the advantage of members and combined meetings are also held, whilst specialised organisations are affiliated to the Society. Together with other transport societies, the Tramway & Light Railway Society is represented on the Consultative Panel for the preservation of British transport relics which has the object of advising the Curator of Historic Relics at the Museum of British Transport. Any enthusiastic tramway modeller will find a wealth of information available to him by joining this Society; in addition to this a wide range of scale model tramcar castings is available from the Engineering Secretary.

For the modeller who wishes to study actual tramcars in operation and acquire actual prototype information for his model the Tramway Museum Society exists at Crich, near Matlock, Derbyshire. This Society was founded in 1955 and acquired the extensive present site of the tramway museum in 1959. A dedicated band of tramway enthusiasts transformed the terrain into a thriv-

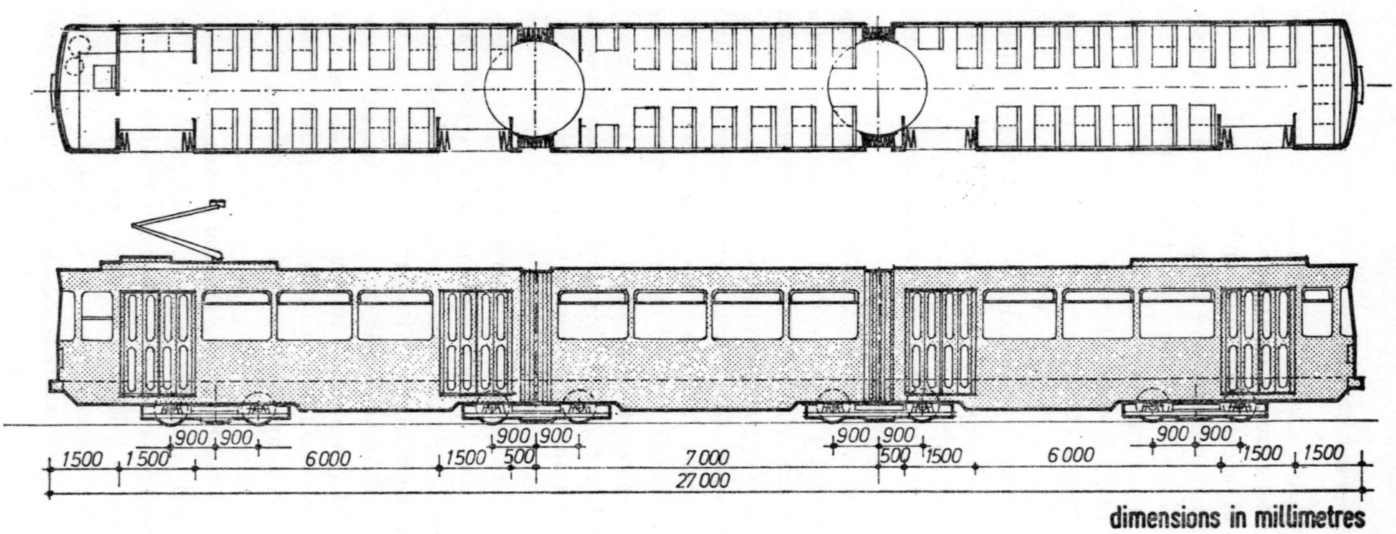

FIG 55 A double-articulated Continental-type rapid-transit railcar operating on the modernised European tramway systems. Double, or single, articulated models of this type are favoured by modellers who have seen them in operation abroad. Drawings can be obtained from a number of European builders

ing mecca for all those interested in electric tramways. Over the years more than forty tramcars have been acquired, depots have been built, and tremendous efforts have resulted in laying sufficient trackwork, erecting the necessary overhead wires and, in 1964, after the installation of a power supply the line was opened for passenger transport. The Society depends entirely on voluntary contributions and, as it is so easily accessible in the centre of the country, carries over 120,000 passengers a year, the crowded car-parks at weekends being impressive evidence of its popularity.

The tramcars owned by the Society are representative of England, Scotland, Ireland, Wales, and a few from overseas countries and any tramcar modeller who visits Crich with a view to undertaking research on the particular type of tramcar he is modelling is bound to find, amongst the dedicated enthusiasts who drive and operate the tramcars, several who have exhaustive knowledge of the tramcar or system he is modelling. Although professional drawings and photographs give invaluable information to a modeller, there really is no substitute for checking the details of, say, the exact shade of the liveries or the distinctive colours of the trucks and lifeguards and, indeed, many other finer points by a visit to a prototype tramcar. These enthusiasts at Crich museum have acquired their fleet of tramcars from the cities and towns where they formerly ran and they have undertaken considerable research themselves to ensure that all details of the restored and rehabilitated tramcars are perfectly accurate. The Tramway Museum Society also publishes *The Journal* a magazine which contains items of interest concerning the maintenance of the museum's tramcar fleet and tramway and gives news of special tramcar events which are constantly being held at Crich.

Another national organisation which the tramway modeller will find of considerable assistance in adding to his knowledge and experience is the Light Railway Transport League which was founded in 1937. The League publishes monthly the magazine *Modern Tramway & Light Railway Review* devoted to descriptions of every aspect of tramways throughout the world in extraordinarily well-informed articles and news items.

For the tramway modeller, particularly those intent on modelling one of the present up-to-date and modern foreign tramway systems, this monthly magazine will prove of inestimable value. He will also find all recently published electric tramway books and literature reviewed as they are issued. The League maintains contact with a wide range of affiliated societies throughout the world which constantly provide it with a wealth of information concerning electric tramways and light rapid-transit undertakings operating in other countries. Regular meetings of members are held in Birmingham, Leeds, Liverpool, London, Manchester, Sheffield, and elsewhere at which interesting talks and lectures on tramways are given. Overseas tramway tours are arranged and these package tours for enthusiasts undertake a previously well-organised itinerary which includes visits to tramway systems and depots over a wide area. The League also has an extensive tramway library service available to members and, also, many special publications are to be had at concession prices.

The Scottish Tramway Museum Society was founded in 1951 and holds regular meetings in Glasgow and Edinburgh. Its illustrated magazine *Scottish Tramlines* is issued free quarterly to members.

In addition to these national organisations there are numerous 'area' clubs which organise local exhibitions, social evenings, lectures and talks, discussions, and facilities for the modeller to perform his lathe work and similar model engineering tasks. They exist in most districts throughout the country and it is unlikely that any modeller, if he makes inquiries, will find himself far from such a club. A number of clubs own tracks available for the use of members while many clubs have individual members owning layouts on which they are only too willing to allow fellow-members to run their models and exchange information on a wide range of matters in connection with the hobby.

Although not a model tramway society, an organisation of considerable help to model tramway engineers is the Society of Model and Experimental Engineers with headquarters at 28 Wanless Road, London. This Society, which was founded by the late Mr Percival Marshall (who devoted a lifetime to model engineering) is probably the oldest and largest in the world in the field of model and light engineering. With the aim of bringing together those who are interested in every aspect of the hobby it encourages craftsmanship through model engineering. It numbers amongst its members many of the finest craftsmen and also beginners and no one can fail to benefit from the advice and assistance available regarding even the most unusual aspect of the hobby. The wide experience and interest of its members and the unequalled resources it possesses appeal alike to all classes of model engineers.

In addition to regular lectures, demonstrations, and visits, the Society has well-equipped workshops, a library and a test room for electrical models. The workshop equipment includes those items which are not usually found in a home workshop but which the tramway modeller uses only occasionally. The Society publishes a well illustrated journal and a monthly news letter, both circulated free to members. Visitors are always welcome when the Headquarters are open and any tramway modeller in the London area would do well to contact the Secretary of this Society.

There are, of course, similar Model and Experimental Engineering Societies in existence in most large cities, which have workshops available and welcome model engineers, to whatever aspect of model engineering they are drawn. Many model tramway engineers have found their workshops invaluable for wheel-turning and similar lathe work.

GLOSSARY OF TRAMWAY NOMENCLATURE

Articulated car	Having a bogie, or bogies common to one or more bodies
Axle	Spindle on or with which wheel revolves
Axlebox	Journal-box in which axle revolves
Amperage	Amount or quantity of electric current
Armature	Piece of soft iron placed in contact with poles of magnet; also the revolving member of an electric motor
Bevel-drive	With cogs oblique to axis
Bogie	Undercarriage pivoted below tramcar
Bonding	Soldering or welding joints in track
Bow-collector	Loop-type current collector on roof of car
Brush	Graphite block in contact with commutator of motor
Bulkhead	Partition between platform and saloon or partitions generally
Canopy-switch	A circuit-breaker installed under the canopy of a car
Cardan-shaft drive	The motor is placed at right-angles to the axle
Check-rails	Inner rails placed inside running rails to assist wheels rounding curves on railway-type track
Controller	Graduated control of current to motors

Crossings	Facing points, when the car runs against the points; trailing points, when the car runs on through the points; crossover, two sets of points enabling crossings from one 'up' line to the 'down' line or vice versa
Destination indicators	Either destination (illuminated) boxes or side destination boards
Disc-wheels	Without spokes
Direct current	Continuous current
Driving wheels	When the axles are motorised
Electro-magnetic brakes	Magnetic track brakes activated by regenerative current from the motors
Fishplates	Plates holding rails together
Gauge	Distance between running rails
Gears	Toothed wheels for transmitting power between motors and axles
Insulator	To isolate opposite polarities of current, or to isolate sections
Lighting-circuit	A term in frequent use to distinguish wiring from 'traction-circuit'
Motorised-axle	Axles which are motor-driven, as distinct from non-motorised axles
Motor-suspension	Method of suspending motors on axles and motor-beams
Notches	The term 'notching-up' is a commonplace tramway term denoting accelerating with the controller
Overhead	Used to denote the trolley-wires to distinguish from conduit (or slot) current collection
Pantograph	A collapsible-type diamond-shaped sprung current collector
Points	Tapering movable rails for directing cars to another line
Rails	Prepared steel (or in modelling brass or aluminium) track
Rail-gauge	A gauge for rail laying

Resistances	Coils of iron ribbon to graduate acceleration
Rolling stock	Used to describe the number of vehicles on a tramway
Roof-planking	The 'catwalk' on the top of double-deck cars to enable engineering staff to attend to trolley-poles
Section-breakers	Insulated blocks inserted in trolley-wires at intervals to sectionalise the current feed
Surface-contact system	An obsolete method of feeding current to the cars by means of energised 'studs' between the running rails
Track-bonding	Soldering cable to each side of joints in track to ensure electrical conductivity
Traction-circuit	The electrical circuit to the motors
Trolley-base	The base of the trolley-pole upon which is fixed the trolley-standard on the roof of the car
Trolley-hangers	The 'ears' used to attach the trolley-wire to the cross-suspension wires
Trolley-head	The container in which the trolley-wheel is inserted
Trolley-pole	This term may refer to the actual trolley-boom on the top of the car or the word 'pole' is also used to describe the street standards supporting the overhead
Voltage-drop	Loss of current due to inadequate supply
Wheel-treads and profiles	Used to describe construction of wheels

ACKNOWLEDGEMENTS

It will be obvious that, in writing a book of this nature, it has been necessary for the author to conduct detailed research in many places, use some ideas which possibly originated with one's friends and be grateful for helpful advice and criticism from them and from firms formerly engaged in the manufacture of electric tramcars and tramway equipment.

In gratefully acknowledging the receipt of such help from the following, I apologise in advance for any which may have been missed out inadvertently from the list: Geoffrey B. Clayton, B. A. Connelly, E. L. Dellow, Gordon Douglas, Richard Elliott, J. C. Franklin, L. G. R. Gibbs, T. A. Gibbs, Peter Hammond, W. J. Haynes, N. Jackson-Stevens, E. R. Oakley, J. R. Prentice, Leo Taylor, Eric Thornton, the late Frank J. Roche and Frank E. Wilson.

My gratitude for technical information is also due to Edgar Allen Engineering Ltd, Brush Electrical Engineering Co Ltd, English Electric-AEI Traction Ltd, Glasgow Corporation Transport Department, London Transport Executive, Maley & Taunton Ltd, Mather & Platt Ltd, The Greater London Council, The Tramway & Light Railway Society, The Tramway Museum Society, The Tramway & Railway World, Model and Allied Publications Ltd.

ILLUSTRATIONS

The sources of illustrations are listed on page 7 and I wish to acknowledge the kindness of individuals and firms who have allowed them to be used in this book. All line drawings are by Mr Eric Thornton with the exception of the two giving standard scale wheel details and also casting details which are the Tramway & Light Railway Society official drawings, whose great assistance I acknowledge.

BIBLIOGRAPHY

Over the past years quite a number of books have been published which will be of interest to the reader and I can recommend the following:

Appleby, J. B., *Bristol Trams Remembered* (J. B. Appleby 1969)

Brook, Roy, *The Tramways of Huddersfield* (Advertiser Press 1959)

Dunbar, C. S., *Buses, Trolleys and Trams* (Paul Hamlyn 1967)

Hearse, G. S., *Tramways of the City of Carlisle* (G. S. Hearse 1962); *The Tramways of Northumberland* (G. S. Hearse 1961)

Hyde, W. G. S., *Manchester Tramway Album* (Manchester Transport Museum Society 1967)

Hardy, P. L., & Jaques, P., *A Short History of Birmingham Corporation Tramways* (H. J. Publications 1970)

Jackson-Stevens, E., *British Electric Tramways* (David & Charles 1971)

'Kennington', *London County Council Tramways Handbook* (Tramway & Light Railway Society 1970)

Klapper, C., *The Golden Age of Tramways* (Routledge & Kegan Paul 1961)

Palmer, G. S., & Turner, B. R., *Blackpool by Tram* (Palmer & Turner 1968)

Pearson, F. K., *Isle of Man Tramways* (David & Charles 1970)

Yearsley, Ian, *The Manchester Tram* (Advertiser Press 1962)

A considerable number of booklets have also been published, in recent years, dealing with specific local electric tramway systems in detail and containing illustrations of the relative systems. All the foregoing books and booklets are obtainable from the Sales Officer of the Tramway & Light Railway Society or from their bookstands at Exhibitions. For those wishing to obtain information about modern foreign tramway systems a profusion of continental books and periodicals are regularly published.

INDEX